Clean Energy Systems and Experiences

Clean Energy Systems and Experiences

Editor

Isak Olsen

scitus academics

Clean Energy Systems and Experiences
Edited by **Isak Olsen**

ISBN: 978-1-68117-295-8
Library of Congress Control Number: 2016937480

SCITUS Academics LLC,
www.scitusacademics.com
Box No. 4766, 616 Corporate Way,
Suite 2, Valley Cottage,
NY 10989

Printed in United States of America on Acid Free Paper

Preface

Clean energy is heat and electricity produced from renewable sources, generating little or no pollution or emissions. These technologies provide clean, renewable sources of power from local sources of energy, which are sustainable over time rather than finite sources like traditional fossil fuels. Nearly every aspect of our daily lives – heating, lighting, cooking, communications, transportation, commerce – depends on electricity. Unfortunately, generating that electricity through the use of traditional fossil fuels can be economically and environmentally costly and a danger to public health. Electricity generated with fossil fuels, such as oil, coal or natural gas, releases into our air and waterways harmful particles into our air, including carbon dioxide, nitrogen oxides, sulfur dioxide, methane and mercury compounds. The clean energy industry generates hundreds of billions in economic activity, and is expected to continue to grow rapidly in the coming years. There's a better, cleaner way to meet our energy needs. Renewable energy resources like wind and solar power generate electricity with little or no pollution and global warming emissions. Renewable energy resources exist over wide geographical areas, in contrast to other energy sources, which are concentrated in a limited number of countries. Rapid deployment of renewable energy and energy efficiency is resulting in significant energy security, climate change mitigation, and economic benefits. This book emphasizes on the latest developments and trends in "clean energy systems and experiences".

Table of Contents

CHAPTER 1

Impact of the Clean Development Mechanism on wind energy investments in Turkey

Murat Tunç and Ruhan Pak

Systems Engineering Department, Engineering Faculty, Yeditepe University, Kayisdagi Cad, Istanbul 34755, Turkey

ABSTRACT

Background

As carbon trading continues to be implemented on both a national and an international scale, it is becoming an important factor in renewable energy investment decisions. Turkey, with continuous growth of carbon dioxide emission and energy consumption since 2001, ratified the Kyoto Protocol in 2009 and began registration of projects with greenhouse gas reductions in 2010. In light of these developments, wind energy resources with a potential of 48,000 MW are among the most efficient and effective solutions for clean and sustainable energy in Turkey. The aim of our study is to reveal the importance of the Clean Development Mechanism (CDM) of the Kyoto Protocol on wind energy investment decisions.

Methods

A broad review of wind energy in Turkey is given, and then, a comprehensive feasibility study of a wind energy firm with a valuation model including Certified Emission Reduction (CER) prices is applied to a case study, the Mega Metallurgy Power. With a holistic and interdisciplinary system engineering approach, results are obtained using comprehensive analysis of technology, emission, and power generation of a wind energy firm linked to a valuation model.

Results

This comprehensive model sets the investment decision-making criteria, the enterprise value comparison with total financing. Finally, a sensitivity analysis is run to show that the enterprise value is positively correlated with CER prices.

Conclusions

Based on these results, it is concluded that if the world's largest carbon offsetting program, the CDM, prevails after 2012, CER prices will have a positive impact on wind energy firm valuations and related investment decisions.

KEYWORDS

Wind energy, Clean development mechanism, Valuation

1. BACKGROUND

1.1 Clean development mechanism

In 2005, the Kyoto Protocol went into effect. It required a strong binding commitment from developed countries to reduce their greenhouse gas emissions from initial emission levels in 1990 to individually assigned levels through the first 5-year commitment period, 2008 to 2012. The Kyoto Protocol provides three flexible mechanisms to help reduce the costs of achieving these emission reduction targets - Emissions Trading, Joint Implementation, and the Clean Development Mechanism (CDM). These mechanisms enable the signatory countries to purchase emission credits from countries that bear lower greenhouse gas reduction costs, without reducing greenhouse gas emissions domestically. The appeal of carbon trading is gaining efficiency by lowering the cost of climate mitigation, which can be accomplished by allowing the market to find the least expensive sources of reduction. CDM is the first and only setup that allows developing countries to participate in reducing greenhouse gases under the United Nations Framework Convention on Climate Change (UNFCCC). The use of offsets is controversial and has limits. Offset usage in trading schemes is usually limited to a proportion of the overall emission target. This limitation ensures that emitters substantially control their own emissions and do not just buy their way out of their obligations. CDM basically relies on three conditions to work: accurate measurement of existing and future emissions, all nations committing to emission reduction targets, and proper verification of carbon offset projects. The CDM and European Union Emissions Trading Scheme (EU ETS) are as yet the only two substantial, legally binding carbon markets in full operation. The EU ETS is the only major scheme of its type in operation, although others are under consideration in the USA, Canada, and Australia. New Zealand began a transitional phase of its ETS in 2010. The EU ETS has completed its trial phase and has begun a second phase concurrent with the Kyoto commitment period. Around 1.3 billion tons of emission savings are expected by the end of that period in 2012. Savings are monitored and verified according to rules set by the United Nations[1].

On the other hand, voluntary markets that mainly deal in offsets attract criticism for lack of verification of actual emission cuts. These markets cater to companies, organizations, and individuals who decide of their own accord to offset their emissions. Because voluntary markets are not part of a compulsory scheme imposed by governments, generally, no authority regulates standards for projects generating carbon offset credits. It is clear that in some recent cases, firms offering carbon credits have accepted money for credits for dubious emission reductions. However, reputable international accreditation schemes backed by international aid groups and environmental organizations have raised the bar in the voluntary market since 2007. Voluntary buyers of offset credits should only consider credits from projects approved by one or more of the carbon offset standards, such as the Gold Standard and Voluntary Carbon Standard. These provide independent third-party verification of emission reductions. By 2008, 96% of offset projects were registered with one or more of these independent standards[2]. In 2009, the global carbon market grew to $144 billion, which has gone up 6% from 2008 despite enduring its most challenging year to date. The voluntary market, which is restricted to 0.2% of the global market, grew to $338 million[3].

A new international framework is still under negotiation and must be established to ensure continuing efforts to reduce global greenhouse gases. To date, however, no solid and tangible agreement has emerged on which to build a post 2012 framework. The recent Copenhagen Accord of 2009 failed to lead to a specific international mechanism to replace the CDM. It seems likely that the CDM - possibly with minor reform - will be part of any new agreement.

1.2 Turkey's energy outlook

Turkey's demand for energy and electricity is increasing rapidly. It is heavily dependent on expensive imported energy resources that burden the economy. The most important renewable resources are hydropower, biomass, geothermal, solar, and wind. Turkey's geographic location has several advantages for extensive use of most of these renewable energy sources. Its hydroelectric potential can meet 33% to 46% of its electric energy demand in 2020. Between 2000 and 2009, power and energy demand in the country have grown at average rates of 4.7% and 5.1%, respectively. Turkey's average annual growth in both power (Figure1) and energy (Figure2) demands between 2010 and 2019 is projected to be 7.2%.

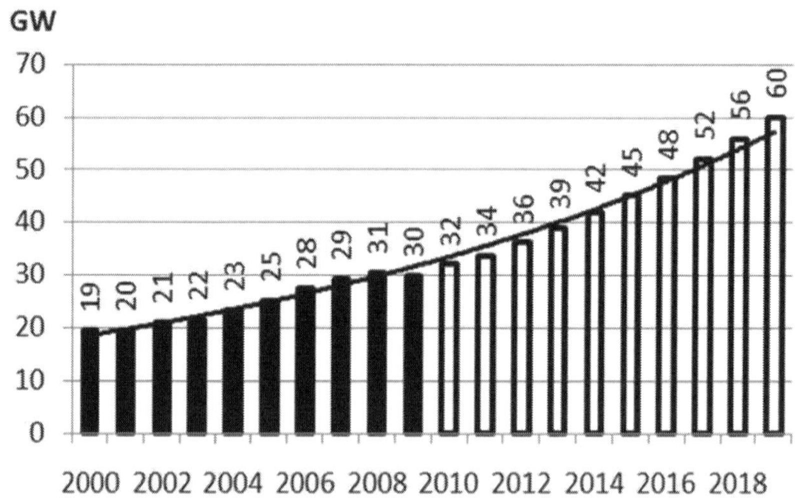

Figure 1 Actual (2000 to 2009) and projected (2010 to 2019) power demand in Turkey (including all inefficiencies).

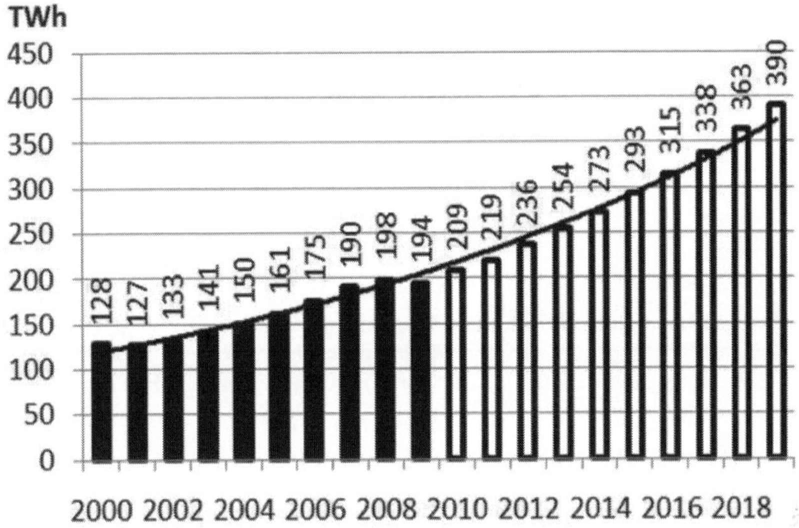

Figure 2 Actual (2000 to 2009) and projected (2010 to 2019) energy demand in Turkey.

Under this scenario, according to the Republic of Turkey Ministry of Energy and Natural Resources, beginning in 2018 at the latest, there will be a supply shortage. This will be so even if the 13,762-MW private and 3,475-MW public projects under construction after 2009 are completed. From 2002 to 2009, natural gas, coal, and hydro resources have been utilized at a growing rate. According to the Energy Market Regulatory Authority, from 2010 to 2015, capacity utilization of these resources will continue to grow. For the same period, a significant power capacity increase is expected for wind, but not for lignite[4].

Regulation, technology, and the political outlook have mainly accounted for environmental and energy policy outcomes in Turkey. The energy policies set by successive governments were highly sensitive to international forces and dynamics. These policies have consistently underplayed the importance of renewable energy and have instead focused on hydropower and nuclear power, despite nationwide protests. In absolute figures, greenhouse gas emissions in Turkey do not seem alarming. However, energy-related carbon emissions have been growing much faster than the economy, at a rate of 6% annually since 1990, and the carbon intensity of the economy is higher than that of developed countries. Therefore, it is inevitable that carbon emissions of Turkey are considered in foreign policy. After ratifying both the UNFCCC in 2004 with differences from Annex I countries in 2001 and the Kyoto Protocol of the Convention in 2009 with Annex I country status, Turkey has become eligible for trading carbon credits under the provisions of the CDM. Core principles of the UNFCCC include that of protecting the climate on an equal basis and with joint effort but different responsibilities. These principles consider needs and specific conditions of developing countries affected by climate change and take effective measures with minimum cost to prevent climate change. Moreover, these principles encourage sustainable development by integrating policies and measures within national development plans. By 2012, Turkey will implement clean development goals to develop institutional capacities and information systems that comply with the provisions of the Convention on an equitable basis. Unlike domestic energy procurement strategies, the global warming dimension of energy politics receives scant attention from civil society and environmental/nongovernmental organizations. A number of promising steps have been taken, such as the Electricity Market Act and Natural Gas Market Law of 2001, toward creating more private involvement and competition. Energy development in Turkey has been dominated by public investment and management. Turkey has made extensive use of financing models such as build-own-operate and build-own-transfer for liberalization, restructuring, and privatization of the energy sector. Although there is steady increase in the use of renewable energy sources, it is still inadequate. Therefore, Turkey needs to focus on governmental grants, feed-in tariffs, and renewable energy certificates[5].

Furthermore, from production to energy use, Turkey needs to increase energy efficiency, prevent waste, and diversify energy supply so that risks associated with a 73% imported energy supply can be mitigated. The country is estimated to have an energy-saving potential of 30% in the construction industry, 20% in the production industry, and 15% in the transportation industry. The Energy Efficiency Law of 2007 and Regulation on Increasing Efficiency in the Use of Energy Sources and Energy of 2008 were enacted to eliminate these inefficiencies. In the course of Turkey's EU accession process, new power generation plants were built in compliance with the Environmental Impact Assessment Regulation and Regulation on Protection of Air Quality from industrial plants, and emission readings from power plants are regularly checked. Reducing electricity generation, transmission, and distribution losses will help reduce emissions. Transmission losses are about 2.5% to 3%, which is within world standards. On the other hand, distribution losses are high. The government aims to reduce losses by privatizing electricity distribution activities. The average ratio of lost and illegally consumed electricity within the distribution system in Organisation for Economic Co-operation and Development (OECD) countries is 4%. This percentage is 6% in the world and around 15% in Turkey. Because it produces more than 80% of global greenhouse gas emissions, the energy sector is key to accomplishing climate change policies[6].

1.3 Wind energy in Turkey

A rapid increase in the utilization of wind energy within the past 15 years was made possible to a large extent by technological developments and a favorable political climate. Since there have been continued improvements in the efficiency and economic competitiveness of wind energy systems, political aspects are now becoming more important. Among them are integration into the national and international power grid, integration into the international energy economy, and societal consensus concerning energy policy. Power generation from wind energy is thus in transition from an alternative to mainstream energy source. Wind can make a decisive future contribution to a climate-compatible and economically feasible power generation system[7]. Among renewable power plants, wind has the major advantage of quick installation, and the disadvantages of supply–demand mismatch, critical efficiency parameters (wind, turbine, and topography), grid integration of dispersed and intermittent wind generation, and noise[8]. The global annual wind resource energy and total installed wind energy power are estimated at 53 TW h and 40,301 MW, respectively. One-third of the latter figure is in Germany. The investment necessary for reaching the world wind energy target of 1,245 GW by 2020 is $900 billion. By then, production costs are expected to decrease from 4.93 to 3.19 cents/kW h (USD) as the annual global business volume in wind turbines increases from $10.4 billion to $104 billion.

According to the Turkish Wind Energy Potential Atlas developed in 2007, the country has a minimum wind energy potential of 5,000 MW in regions with annual wind speeds of 8.5 m/s and higher, and 48,000 MW in areas with speeds higher than 7.0 m/s. There has been some progress in increasing the amount of installed wind energy power, which was only 18 MW as of 2004. By the end of 2009, this figure reached 802.8 MW. Upon enactment of the Renewable Energy Law, licenses were granted to 93 new wind projects that deliver a total installed power of 3,363 MW. Out of these projects, 1,100 MW were under construction in 2010[6]. The Renewable Energy Law of 2005 also introduced the government's purchasing guarantee for electricity procured from renewable resources and hence paved the way for wind energy investments. Within the same framework, sales of renewable energy in the free market were defined. Besides such legal arrangements, the government still needs to have the transmission system ameliorated, bureaucratic studies coordinated, and incentives for promoting wind energy resolved. On the other hand, all wind power plants are owned by investors from the private sector, and these investors need to focus more on complying with grid connection criteria.

When the distribution of wind potential by geographic regions is evaluated, wind speed is generally found to be 6.0 to 7.0 m/s on the Marmara, western Black Sea, and eastern Mediterranean coasts; 5.0 to 6.0 m/s on the western

Mediterranean coasts; 7.0 to 8.5 m/s on the north western Aegean coasts; and 4.5 to 5.5 m/s in interior regions[9].

Eventually, firms willing to exploit this considerable wind energy potential must take carbon trading into consideration in their investment decisions[10]. This development raises a question: what will be the impact of the CDM on valuations of wind energy firms?

2. METHODS

2.1 Case study: mega metallurgy power

To demonstrate the impact of the CDM on valuations of wind energy firms, we investigate a case study, Mega Metallurgy Power. There is no prior study that links a comprehensive analysis of technology, emission, and power generation of a wind energy firm with a valuation model that includes Certified Emission Reduction (CER) prices.

Mega Metallurgy is a steel production firm in the Payas Organized Industrial Zone/Hatay, with an annual capacity of 50,000 tons. Mega Metallurgy seeks permanent solutions, such as a wind energy project in Hatay, instead of temporary ones, like building a de-dusting plant to decrease its air pollution. Establishing a wind energy firm (Mega Metallurgy Power) may enable Mega Metallurgy to offset its own air pollution, averting the need for CER purchases or the probable incurrence of EU penalties after 2012. To make this investment decision, a valuation model including the CDM, along with analysis of CER prices on valuations, is required.

Technology analysis is carried out to determine the most eligible wind turbine at the time of investment to benefit from the continuous progress in wind energy systems. Once the wind turbines are selected, the emission analysis is conducted to find annual allowed CO_2 emissions. The power analysis is next, with which the required number of turbines and their installed power capacity and CO_2 emission savings are calculated, according to the selected wind turbines and annual allowed CO_2 emissions. Power generation and CO_2 emission saving data are fed into the financial analysis, from which the Mega Metallurgy Power enterprise value is derived, taking the CDM into account. This comprehensive model sets the investment decision-making criteria, which is the enterprise value comparison with total financing. Finally, a sensitivity analysis is run to conclude the positive correlation between the CER price and enterprise value.

With a holistic and interdisciplinary system engineering approach, the initial groundwork for systematic analysis of wind energy investments in Turkey and abroad, wherever the CDM is in effect, is laid in this study.

3. RESULTS AND DISCUSSION

3.1 Technology analysis

Wind turbine manufacturers are turning away from industry standard gearboxes and generators, in a bid to boost reliability and reduce the cost of wind power. Siemens has begun selling a 3-MW turbine with a direct drive system that replaces the conventional high-speed generator with a low-speed one, which eliminates the need for a gearbox. Recently, General Electric (GE) announced an investment of $442 million in manufacturing facilities to build its own 4-MW direct drive turbines for offshore wind farms. Siemens' plans hinge on a new design that reduces the weight of the system generator. In conventional wind turbines, the gearbox increases the speed of the wind-driven rotor several hundredfold, which radically reduces the size of the required generator. Direct drive generators operate at the same speed as the turbine blades and must therefore be much larger - over 4 m in diameter for Siemens' 3-MW turbines. Yet, Siemens claims that the turbine's entire nacelle weighs just 73 metric tons - 12 tons less than its less powerful, gear-driven 2.3-MW turbines. Much of the weight reduction comes from the use of permanent magnets in the generator's rotor - a setup that GE is also using. Conventional turbine generators use electromagnets - copper coils fed with electricity from the generator itself. A 15-mm-thick segment of permanent magnets can generate the same magnetic field as a 10- to 15-cm section of copper coils. Siemens further reduced the weight by inverting its generator design. Rather than a steel rotor covered with permanent magnets spinning inside a stationary doughnut-shaped stator, the Siemens rotor is a steel cylinder with permanent magnets on the inside, and this rotor spins around a column-like stator. Siemens erected a prototype of its machine in Brande, Denmark in December 2010 and plans to begin mass production in 2011. GE's technology, which it acquired with the purchase of the Norwegian turbine producer ScanWind in 2009, is being demonstrated at a test site in Norway. Commercialization of its 4-MW machine is slated for 2012. The direct drive system, with half as many parts as a conventional turbine and no gearbox, increases reliability. However, direct drive systems introduce one potential problem. There are ongoing concerns regarding future supply of the rare earth metals used to make permanent magnets[11]. The first 2.5-MW Goldwind permanent magnet direct drive (PMDD) prototype was installed in China in 2009, and full production was initiated in February 2010. In November 2010, Goldwind's Jiangsu Dafeng offshore turbine plant was founded. Upon completion, it will be the largest offshore wind power equipment manufacturing site in China and one of the largest in the world, with a production capacity of 300 turbines per year. Recently, Goldwind has also developed a 3-MW turbine that was successfully installed and that began trial operation in Dabancheng. During erection and its parts, the 2.5-MW Goldwind PMDD wind turbine is seen in Figures3 and4, respectively.

Figure 3 The 2.5-MW Goldwind PMDD wind turbine during erection.

Mega Metallurgy selected Goldwind's 2.5-MW PMDD wind turbines not only because of their availability and previous installation successes in Turkey, but also because of their technical specifications (Table1) and the following features: automatic lubricating systems, dry-toothed belt pitch drive system, greased gear spurs ensuring fewer maintenance issues, robust cooling system, no intermediate transmission, advanced variable speed control, wide adjustable control range, better quality electrical output, advanced power controls allowing for complete adjustability of total output and rate of change, increased reactive power, adjustable output power, low voltage ride through, and successful grid connection[12].

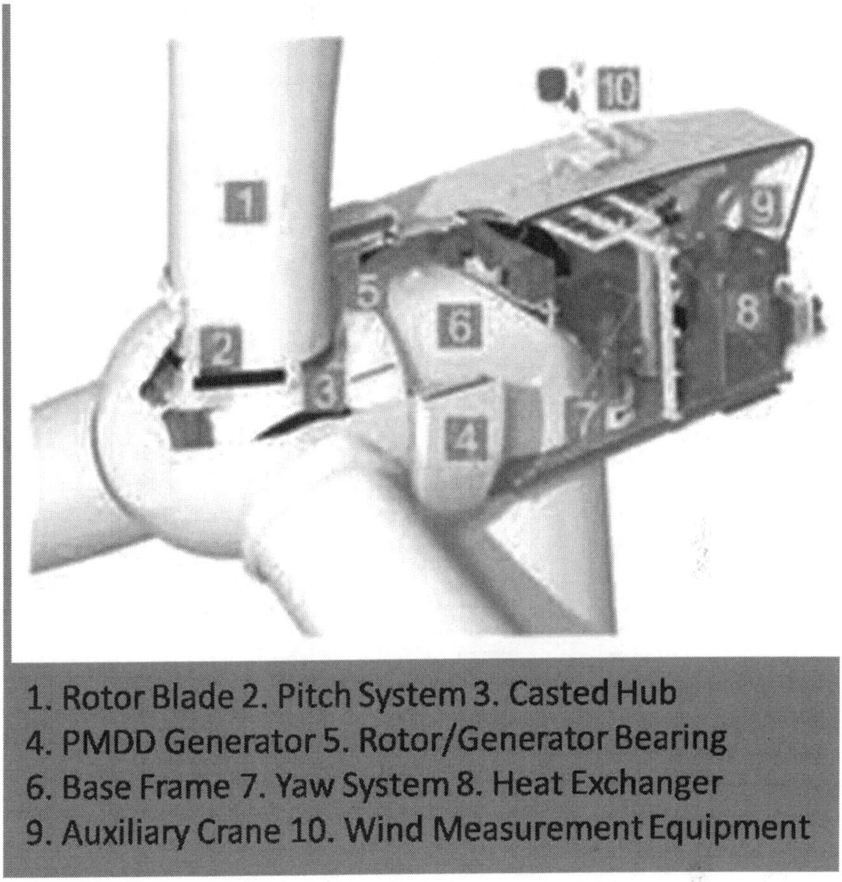

Figure 4 The 2.5-MW Goldwind PMDD wind turbine section.

3.2 Emission analysis

Turkey's CO_2 emission coefficient (C_{TR}), which is defined as the number of tons of CO_2 emitted per gigawatt hour of energy produced, is 0.511. This coefficient is calculated by dividing the 101,473,200 tons of annual CO_2 emissions of all power plants in the country (X_{P08})[13] by the 198,418,000 GW h of annual electricity production of all power plants in 2008 (E_{P08})[14] (Equation 1):

$$C_{TR} = \frac{X_{P08}}{E_{P08}} \tag{1}$$

Table 1 Technical specifications of Goldwind's 2.5-MW turbine

Specification	**Value**
Rated power (KW)	2,500
Rotor diameter (m)	100
Swept area (m^2)	7,854
Cut-in wind speed (m/s)	3
Cut-out wind speed (m/s)	20
Hub height (m)	80
Power coefficient (C_p)	0.47
Number of blades	3

In 2008, Mega Metallurgy, with an annual production of 50,000 tons of steel (M), is assumed to operate at the same 2010 CO_2 emission level as Isdemir[15], which is a subsidiary of Erdemir, the largest steel production company in the country. Therefore, the CO_2 emission coefficient of Mega Metallurgy per crude steel production (C_M) is 2.05. The total annual CO_2 emission of Mega Metallurgy (X_{M08}) is calculated at 102,500 tons (Equation 2):

$$X_{M08} = MC_M \tag{2}$$

Because the plant was not operational in 1990, the national steel industry CO_2 emission inventory data of 9,402,020 tons (X_{SI90}) for 1990[16] and 12,769,350 tons (X_{SI08}) for 2008[13] are proportioned, to find the 1990 CO_2 emission of Mega Metallurgy (X_{M90}), which is 75,470 tons (Equation 3):

$$X_{M90} = \frac{X_{SI90} X_{M08}}{X_{SI08}} \tag{3}$$

To comply with the Kyoto Protocol, the plant should decrease its CO_2 emissions by 5.2% relative to 1990 levels. Therefore, allowed annual CO_2 emission for Mega Metallurgy (X_{MA}) is 71,550 tons (Equation 4), and hence, the firm should make up for the excess 30,950 tons (X_{ME}) (Equation 5).

$$X_{MA} = X_{M90}(1 - 0.052) \tag{4}$$

$$X_{ME} = X_{M08} - X_{MA} \tag{5}$$

Otherwise, the firm would have to buy CERs at market rate. The prices of CERs fluctuate primarily because of the costs of projects that generate them and supply–demand fundamentals that drive their trade on financial markets. Like energy prices, carbon prices are very sensitive to energy supply shocks. The escalating international oil price also feeds higher prices for carbon emission permits or EU allowances. Because of rising gas prices, dirtier coal becomes a more attractive fuel for EU power generators and other heavy users of fossil fuels covered by the EU ETS, such as the iron and steel industries. The price may increase further as the market digests a new emission outlook in Europe. When an entity covered by the ETS misses the deadlines for surrendering emission allowances, it runs the risk of triggering enforcement procedures. The EU ETS Directive calls for a $130 penalty for each ton of carbon dioxide equivalent emitted, for which the operator has not surrendered allowances[17]. Phase 3 of the EU ETS could see the price of EU allowances (EUA) rise from $39/ton in 2013 to $52/ton in 2016. Such an increase, from current levels of around $19.5/ton, is vital to meet the EU's current emission reduction targets of 20% below 1990 levels. There are three likely scenarios. First, there is the EU-20 scenario, whereby the EU does not increase its current emission reduction target. Second, there is the EU-30 scenario, where the current reduction target is increased to 30% below 1990 levels, and third, there is the ETS linking scenario, in which the EU ETS establishes a full link with the US ETS and other ETS schemes before 2018. The EU-20 scenario would lead to a price range of $32.5 to $65/ton by 2016; the EU-30 scenario, to a price range of $45.5 to $84.5/ton by 2016; and the final scenario, to a price range of $13 to $39/ton by 2016. The latter is significantly lower than the first two scenarios because of new and expanded crediting mechanisms and a larger import limit. In 2009, emission projections were dramatically reduced following the economic downturn, which resulted in lower emissions and a surplus of EUA and CERs in phase 2 that can be banked into phase 3. At the same time, the USA has become an active and constructive participant in international climate negotiations, and a climate agreement seems within reach. Therefore, it is now more likely that the EU will adopt a 30% reduction target. Furthermore, cap-and-trade schemes in the USA and other regions have become more likely, which increase the potential of the USA linking with the EU ETS[18].

3.3 Power generation analysis

Once an eligible location in Hatay was selected, wind velocities at the following coordinates (Table2) were measured monthly for 6 years (Table3) and daily (Table4) using masts.

Table 2 Turbine coordinates

Coordinates (N)	Coordinates (W)	Altitude (m)
36°28/57"	36°12/37"	678
36°28/44"	26°12/32"	806
36°28/39"	36°13/03"	761
36°28/34"	36°12/38"	810
36°28/59"	36°13/02"	745

Table 3 Wind velocity measurements (m/s)

Month	Years						
	2004	2005	2006	2007	2008	2009	Mean
January	6.91	6.81	6.91	6.70	6.86	7.02	6.87
February	6.89	6.91	6.96	6.36	6.62	6.88	6.77
March	6.71	6.50	6.13	6.25	6.26	6.53	6.40
April	5.60	6.26	6.00	6.39	6.69	3.85	5.80
May	5.67	5.95	5.42	5.63	5.37	6.99	5.84
June	5.60	5.76	6.08	6.26	6.65	6.57	6.15
July	6.51	6.83	4.81	6.62	6.31	6.13	6.20
August	6.00	6.79	6.16	6.20	6.17	6.01	6.22
September	6.75	6.68	6.49	6.30	6.16	6.36	6.46
October	7.58	6.61	6.37	6.95	6.82	6.68	6.84
November	6.81	6.97	7.03	6.90	6.60	7.01	6.89
December	7.76	7.03	7.58	7.21	7.68	7.67	7.49
Mean	6.57	6.59	6.33	6.48	6.52	6.48	6.49

Table 4 Daily wind velocity time series

	Hour																							
	1	2	3	4	5	6	7	8	9	10	11	12	13	14	15	16	17	18	19	20	21	22	23	24
Velocity (m/s)	3.3	3.8	4.2	3.3	2.8	3.0	4.0	2.7	5.2	6.7	6.8	6.8	5.7	8.3	9.2	9.3	6.5	4.2	4.3	3.7	4.0	2.8	3.7	3.3

Then, on the basis of a Weibull distribution, the cumulative frequency of wind velocities between cut-in and cut-out limits of the turbine is calculated (Table5).

Table 5 Daily wind velocity frequency distribution (Weibull distribution)

	Velocity (m/s)									
	(0–1)	(1–2)	(2–3)	(3–4)	(4–5)	(5–6)	(6–7)	(7–8)	(8–9)	(9–10)
Frequency	2%	7%	9%	15%	20%	17%	8%	1%	0%	0%

The capacity factor (C_W) of the location is calculated at 40% (Table6).

Table 6 Monthly capacity factors

	Months												
	Jan	Feb	Mar	Apr	May	Jun	Jul	Aug	Sep	Oct	Nov	Dec	Mean
Rates	45%	48%	41%	30%	31%	36%	37%	37%	42%	39%	40%	48%	40%

Wind turbine power (P_W) is calculated using the following: 7,857 m^2 as the area swept by the blades (A), 1.225 kg/m^3 as air density at 15°C and normal atmospheric pressure (ρ), a 0.47 turbine power coefficient (C_P), and 6.49 m/s as wind velocity (V). This calculation (Equation 6) yields a turbine power of 619 kW:

$$P_W = 0.5 C_P \rho A V^3 \quad (6)$$

Annual capacity (C) is calculated at 3,460 h (Equation 7):

$$C = 8,760 C_W \quad (7)$$

The annual energy (E_W) per turbine is calculated at 2,142,344 kW h (Equation 8):

$$E_W = P_W C \qquad (8)$$

Annual CO_2 savings (S) per turbine is 1,096 tons (Equation 9):

$$S = \frac{C_{TR} E_W}{1,000} \qquad (9)$$

The number of turbines required (N_T) is 28 (Equation 10), and hence, the total installed capacity (P_{WT}) is 70 MW (Equation 11) with an estimated annual energy production (E_{WT}) of 59,985,643 kW h (Equation 12) and total annual CO_2 savings (S_T) of 30,667 tons (Equation 13).

$$N_T = \frac{X_{ME}}{S} \qquad (10)$$

$$P_{WT} = 2,5 N_T \qquad (11)$$

$$E_{WT} = E_W N_T \qquad (12)$$

$$S_T = S N_T \qquad (13)$$

At an average CER price estimation (P_{CER}) of $40/ton that is based on future CER price analysis, annual savings ($S_\$$) for establishing this wind power plant would be $1,227,093 (Equation 14).

$$S_{\$} = S_T P_{CER} \tag{14}$$

3.4 Financial analysis

Twenty-eight 2.5-MW Goldwind wind turbines with an 80-m hub height cost $56 million, which includes transportation costs, crane hiring for erection, a supervisory control and data acquisition system for remote monitoring and supervision, and aviation lights. The availability of high leverages in the local financial market for such renewable energy projects is an advantage for Mega Metallurgy. A total of $39.2 million in long-term loan drawdowns with a 10-year term, 1-year grace period, equal principal repayments, semi-annual interest payments at an all-in cost interest rate of 3% with an upfront arrangement fee of 1.5%, and a 70% long-term (L/T) loan/(equity + L/T loan) ratio can be obtained from local creditors (Table7).

Table 7 Mega metallurgy power long-term loan amortization schedule (in dollars)

Drawdowns	Balance	Available line	Date	Days	Repayment	Interest	Fee expense	Cash flow	Period
39,200,000	39,200,000	0	01.01.2012	0	0	0	588,000	−38,612,000	0
0	39,200,000	0	30.06.2012	181	0	591,267		591,267	1
0	37,136,842	2,063,158	31.12.2012	184	2,063,158	601,067		2,664,225	2
0	35,073,684	4,126,316	30.06.2013	181	2,063,158	560,147		2,623,305	3
0	33,010,526	6,189,474	31.12.2013	184	2,063,158	537,796		2,600,954	4
0	30,947,368	8,252,632	30.06.2014	181	2,063,158	497,909		2,561,067	5
0	28,884,211	10,315,789	31.12.2014	184	2,063,158	474,526		2,537,684	6
0	26,821,053	12,378,947	30.06.2015	181	2,063,158	435,670		2,498,828	7
0	24,757,895	14,442,105	31.12.2015	184	2,063,158	411,256		2,474,414	8
0	22,694,737	16,505,263	30.06.2016	182	2,063,158	375,495		2,438,653	9
0	20,631,579	18,568,421	31.12.2016	184	2,063,158	347,986		2,411,144	10
0	18,568,421	20,631,579	30.06.2017	181	2,063,158	311,193		2,374,351	11
0	16,505,263	22,694,737	31.12.2017	184	2,063,158	284,716		2,347,874	12
0	14,442,105	24,757,895	30.06.2018	181	2,063,158	248,954		2,312,112	13
0	12,378,947	26,821,053	31.12.2018	184	2,063,158	221,446		2,284,604	14
0	10,315,789	28,884,211	30.06.2019	181	2,063,158	186,716		2,249,874	15
0	8,252,632	30,947,368	31.12.2019	184	2,063,158	158,175		2,221,333	16
0	6,189,474	33,010,526	30.06.2020	182	2,063,158	125,165		2,188,323	17
0	4,126,316	35,073,684	31.12.2020	184	2,063,158	94,905		2,158,063	18
0	2,063,158	37,136,842	30.06.2021	181	2,063,158	62,239		2,125,396	19
0	0	39,200,000	31.12.2021	184	2,063,158	31,635		2,094,793	20
39,200,000				*Total*	39,200,000	6,558,263	588,000	3.38%	*IRR*

For a 10-year (2012 to 2021) financial projection of Mega Metallurgy Power, interest, foreign exchange, and inflation rates are assumed constant, and cash flow deficits are financed by short-term (S/T) loans (Table8).

Table 8 Mega metallurgy power projected balance sheet (in dollars)

MMP balance sheet ($)	31.12.2012	31.12.2013	31.12.2014	31.12.2015	31.12.2016	31.12.2017	31.12.2018	31.12.2019	31.12.2020	31.12.2021
Accounts receivable	0	364,913	364,913	364,913	364,913	364,913	364,913	364,913	364,913	364,913
Current assets	0	364,913	364,913	364,913	364,913	364,913	364,913	364,913	364,913	364,913
Other fixed assets	1,111,936	1,111,936	1,111,936	1,111,936	1,111,936	1,111,936	1,111,936	1,111,936	1,111,936	1,111,936
Fixed assets	56,000,000	56,000,000	56,000,000	56,000,000	56,000,000	56,000,000	56,000,000	56,000,000	56,000,000	56,000,000
Accumulated depreciation	−2,911,194	−5,822,387	−8,733,581	−1,644,774	−14,555,968	−17,467,162	−20,378,355	−23,289,549	−26,200,742	−29,111,936
Total assets	*54,200,742*	*51,654,461*	*48,743,268*	*45,832,074*	*42,920,881*	*40,009,687*	*37,098,493*	*34,187,300*	*31,276,106*	*28,364,913*
Accounts payable	0	19,167	19,167	19,167	19,167	19,167	19,167	19,167	19,167	19,167
S/T bank loans	5,216,239	5,694,947	5,702,792	5,578,936	5,350,184	5,281,408	5,105,176	4,817,010	4,412,820	3,886,625
L/T fund loans due	4,126,316	4,126,316	4,126,316	4,126,316	4,126,316	4,126,316	4,126,316	4,126,316	4,126,316	0
Current liabilities	9,342,555	9,840,429	9,848,275	9,724,418	9,495,666	9,426,890	9,250,658	8,962,493	8,558,302	3,905,791
L/T fund loans	33,010,526	28,884,211	24,757,895	20,631,579	16,505,263	12,378,947	8,252,632	4,126,316	0	0
Current liabilities	9,342,555	9,840,429	9,848,275	9,724,418	9,495,666	9,426,890	9,250,658	8,962,493	8,558,302	3,905,791
L/T fund loans	33,010,526	28,884,211	24,757,895	20,631,579	16,505,263	12,378,947	8,252,632	4,126,316	0	0
L/T liabilities	33,010,526	28,884,211	24,757,895	20,631,579	16,505,263	12,378,947	8,252,632	4,126,316	0	0
Capital	16,800,000	16,800,000	16,800,000	16,800,000	16,800,000	16,800,000	16,800,000	16,800,000	16,800,000	16,800,000
Retained earnings	0	−4,952,339	−3,870,178	−2,662,901	−1,323,923	119,951	1,403,849	2,795,204	4,298,492	5,917,804
Net income	−4,952,339	1,082,160	1,207,277	1,338,979	1,443,874	1,283,898	1,391,354	1,503,288	1,619,312	1,741,317
Total liabilities and equity	*54,200,742*	*51,654,461*	*48,743,268*	*45,832,074*	*42,920,881*	*40,009,687*	*37,098,493*	*34,187,300*	*31,276,106*	*28,364,913*

Electricity sale price is taken as 7.3 cent/kW h, which is in line with the final draft of the Renewable Energy Law (Table9).

Table 9 Mega metallurgy power projected income table (in dollars)

MMP income table ($)	2012	2013	2014	2015	2016	2017	2018	2019	2020	2021
Sales	0	4,378,952	4,378,952	4,378,952	4,378,952	4,378,952	4,378,952	4,378,952	4,378,952	4,378,952
Net carbon emission savings	0	1,227,093	1,227,093	1,227,093	1,227,093	1,227,093	1,227,093	1,227,093	1,227,093	1,227,093
Operating expense	0	−230,000	−230,000	−230,000	−230,000	−230,000	−230,000	−230,000	−230,000	−230,000
Overhead expense	0	−50,000	−50,000	−50,000	−50,000	−50,000	−50,000	−50,000	−50,000	−50,000
Maintenance expense (Goldwind)	0	−100,000	−100,000	−100,000	−100,000	−100,000	−100,000	−100,000	−100,000	−100,000
Administrative and operational expenses	0	−30,000	−30,000	−30,000	−30,000	−30,000	−30,000	−30,000	−30,000	−30,000
All risk insurance expense	0	−50,000	−50,000	−50,000	−50,000	−50,000	−50,000	−50,000	−50,000	−50,000
Depreciation expense	−2,911,194	−2,911,194	−2,911,194	−2,911,194	−2,911,194	−2,911,194	−2,911,194	−2,911,194	−2,911,194	−2,911,194
Interest expense	−2,041,145	−1,382,691	−1,257,575	−1,125,873	−990,990	−859,979	−725,659	−585,742	−440,711	−288,205
Corporate tax at 20%	0	0	0	0	−29,988	−320,975	−347,839	−375,822	−404,828	−435,329
Net income	*−4,952,339*	*1,082,160*	*1,207,277*	*1,338,979*	*1,443,874*	*1,283,898*	*1,391,354*	*1,503,288*	*1,619,312*	*1,741,317*

Mega Metallurgy Power performance is taken as the market average; hence, its beta is 1. To calculate the cost of debt (K_d), the cost of short- and long-term debts before tax is assumed to be the market rates of 5% and 6%, respectively. When calculating an 11% cost of equity (K_e), the risk-free rate and market risk premium are taken as a 6% yield on 10-year Turkish Eurobonds and 5% according to Turkish sovereign risk, respectively. The corporate tax rate is 20%, and the company distributes no dividends. Company financials are assumed to stabilize in 2017, with 46% weight of equity (W_e) and 8% weighted average cost of capital (WACC). Finally, a discounted cash flow value of $68.6 million for Mega Metallurgy Power is derived at a 1% terminal growth rate[19] (Table10).

Table 10 Mega metallurgy power valuation (in dollars)

MMP discounted cash flow valuation ($)	2012	2013	2014	2015	2016	2017	2018	2019	2020	2021
W_e	22%	25%	29%	34%	39%	46%	53%	62%	73%	86%
W_d	78%	75%	71%	66%	61%	54%	47%	38%	27%	14%
K_d	6%	6%	6%	6%	6%	6%	6%	6%	5%	5%
WACC	*6%*	*6%*	*7%*	*7%*	*7%*	*8%*	*8%*	*9%*	*9%*	*10%*
EBITDA	0	5,376,045	5,376,045	5,376,045	5,376,045	5,376,045	5,376,045	5,376,045	5,376,045	5,376,045
Non-cash working capital change	0	−345,746	0	0	0	0	0	0	0	0
Tax	0	0	0	0	−29,988	−320,975	−347,839	−375,822	−404,828	−435,329
Free cash flow	-	*5,030,299*	*5,376,045*	*5,376,045*	*5,346,057*	*5,055,071*	*5,028,207*	*5,000,223*	*4,971,217*	*4,940,716*
Discount factor	-	1.06	1.13	1.21	1.30	1.39	1.50	1.63	1.78	1.96
Present value	-	*4,733,763*	*4,750,119*	*4,447,802*	*4,128,026*	*3,630,450*	*3,344,706*	*3,065,080*	*2,790,739*	*2,520,549*
Sum of present values	21,690,159									
Terminal growth rate	1%									
Terminal value	78,347,787									
Terminal value present value	56,267,802									
Firm value	*77,957,961*									
Net cash position	*−9,342,555*									
Discounted cash flow value	*68,615,406*									

As for the sensitivity analysis, valuation is calculated for different CER prices in which the only change to company financials is stabilization years. Although the WACC of each simulation must be exactly the same to make a perfect comparison, approximations are used so that full financial years are not split. Consequently, the enterprise value is found positively correlated with the CER prices, as seen in Figure5.

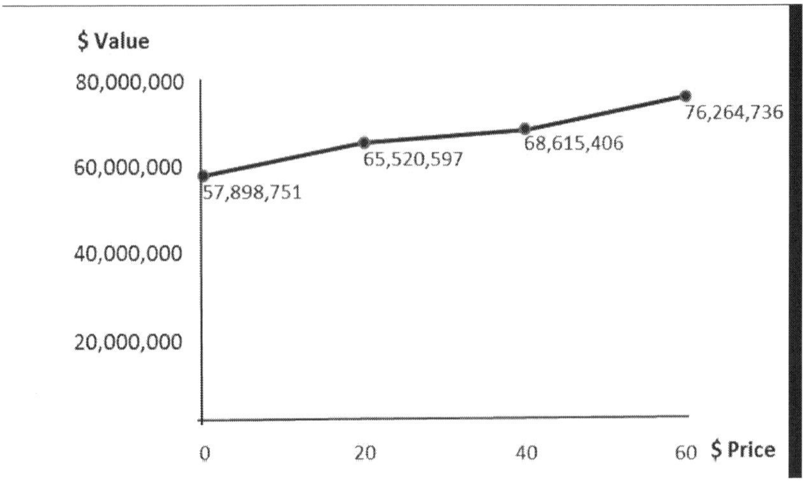

Figure 5 Effect of CER price (dollars per ton) on Mega Metallurgy Power valuations.

4. CONCLUSIONS

If the CDM, the world's largest carbon offsetting program, prevails after 2012, CER prices will continue to have a positive impact on wind energy firm valuations and related investment decisions. However, it is also concluded that in Turkey, CER prices alone would not be sufficient to justify wind energy investment. Wind velocity is still the main determining factor, as in the case of Mega Metallurgy Power. Comparing a total financing of \$61.2 million (\$16.8 million equity + \$39.2 million long-term loan + \$5.2 million short-term loan) with enterprise value creation of \$68.6 million, Mega Metallurgy does not find this project investment appealing. Therefore, Mega Metallurgy should consider investing in a different region with wind velocity high enough to justify their investment. Besides wind velocity and CER prices, future developments in technology and electricity sale prices should also be closely monitored for changes that would have a significant effect on investment decisions.

COMPETING INTERESTS

The authors declare that they have no competing interests.

AUTHORS' CONTRIBUTIONS

MT carried out the emission and power generation analysis, and RP participated in these studies. RP carried out the technology and financial analysis, and MT participated in these studies. RP drafted the manuscript and participated in the design of the study. MT conceived the study, participated in its design, and helped draft the manuscript. All authors read and approved the final manuscript.

REFERENCES

1 UNFCCC: **United Nations framework convention on climate change.** 2011.http://unfccc.int Accessed 26 Dec 2011

2 Hamilton K, Sjardin M: **Fortifying the foundation: state of the voluntary carbon markets 2009.** *Ecosystem marketplace and new carbon finance* 2009. http:// www. ecosystemmarketplace. com/pages/dynamic/resources.library.page.php?page_id=7082& section=our_publications&eod=1 . Accessed 26 Dec 2011

3 Ambrosi P, Kossoy A: **State and trends of the carbon market 2010.** 2010.http://web.worldbank.org/WBSITE/EXTERNAL/TOPICS/ ENVIRONMENT/EXTCARBONFINANCE/0,contentMDK: 22592488~pagePK:64168445~piPK:64168309~theSitePK:4125853,00. htmlAvailable via the World Bank.. Accessed 26 Dec 2011

4 TEIAS: **10-Year electricity production capacity projection of Turkey (2010–2019).** *Turkish Electricity Transmission Company Research Planning and Coordination* 2011. http://www.emra.org.tr/ index.php/electricity-market/electricity-publishments. Accessed 26 Dec 2011

5 Yüksel I: **Energy production and sustainable energy policies in Turkey.** *Renewable Energy* 2010,**35**(7):1469–1476. 10.1016/j.renene.2010.01.013

6 Republic of Turkey Ministry of Energy and Natural Resources: **Wind.** 2010.http://www.enerji.gov.tr/index.php?dil=en&sf= webpages&b=ruzgar_EN&bn=231&hn=&nm=40717&id=40734 . Accessed 26 Dec 2011

7 Kühn M: **A tailwind for sustainable technology.** In *Renewable Energy - Sustainable Concepts for the Future.* Edited by: Bührke T, Wengenmayr R. Wiley-VCH, Weinheim; 2008.

8 Akova I: *Renewable energy resources*. Nobel Yayın Dagitim, Istanbul; 2008.

9 Akova I: **Development potential of wind energy in Turkey.** 2011.http://echogeo.revues.org/12457?lang=en Available via Echogeo.. Accessed 10 July 2012

10 Çamdali U, Parmaksizoğlu C, Tunç M: **Comparison of Turkey's electrical energy consumption and production with some European countries and optimization of future electrical power supply investments in Turkey.** *Energy Policy* 2006, **34**(1):50–59. 10.1016/j.enpol.2004.04.027

11 Fairley P: **Wind turbines shed their gears.** 2010.http://www.technologyreview.com/energy/25188/page1/?a=f Available via technology review: published by MIT.. Accessed 26 Dec 2011

12 Goldwind Science & Technology Co., Ltd: **Goldwind.,.** 2010.http://www.goldwindglobal.com/web/index.do . Accessed 26 Dec 2011

13 UNFCCC: **National inventories submissions 2010 (TUR-2010–2008-v1.2.xls/Table1s1).** *UN framework of climate change* 2010. http://unfccc.int/files/national_reports/annex_i_ghg_inventories/national_inventories_submissions/application/zip/tur-2010-crf-15april.zip. Accessed 26 Dec 2011

14 Turkish Electricity Transmission Company: **Annual development of installed capacity and generation in Turkey.** 2008.http://www.teias.gov.tr/TürkiyeElektrikİstatistikleri/istatistik2008/13.xls . Accessed 26 Dec 2011

15 Isdemir: **Sustainability report.** 2010.http://www.isdemir.com.tr/wps/wcm/connect/isdemirlibrary/isdemireng/sustainability/isdemircontentsustainabilityreport . Accessed 26 Dec 2011

16 UNFCCC: **National inventories submissions 2010 (TUR-2010–1990-v1.2.xls/Table1s1).** *UN framework of climate change* 2010. http://unfccc.int/files/national_reports/annex_i_ghg_inventories/national_inventories_submissions/application/zip/tur-2010-crf-15april.zip. Accessed 26 Dec 2011

17 EUR-Lex: **32003L0087: EU directives 2003/87/EC and 96/61/EC.** 2003.http://eur-lex.europa.eu/LexUriServ/LexUriServ.do?uri=CELEX:32003L0087:EN:HTML . Accessed 26 Dec 2011

18 Renewable Energy World: **Carbon trading price rise predicted.** 2009.http://www.renewableenergyworld.com/rea/news/article/2009/06/carbon-trading-price-rise-predicted . Accessed 26 Dec 2011

19 Rosenbaum J, Pearl J: *Investment banking: valuation, leveraged buyouts, and mergers and acquisitions*. John Wiley & Sons, Inc, Hoboken; 2009.

CHAPTER 2

Cost Benefit Analysis of Using Clean Energy Supplies to Reduce Greenhouse Gas Emissions of Global Automotive Manufacturing

Qiang Zhai [1], Huajun Cao [2], Xiang Zhao [3] and Chris Yuan [1,*]

[1] Department of Mechanical Engineering, University of Wisconsin Milwaukee, Milwaukee, WI 53201, USA

[2] School of Mechanical Engineering, Chongqing University, Chongqing, 400044, China

[3] Manufacturing Systems Research Lab, Global Research and Development, General Motors Company, Warren, MI 48090, USA

ABSTRACT

Automotive manufacturing is energy-intensive. The consumed energy contributes to the generation of significant amounts of greenhouse gas (GHG) emissions by the automotive manufacturing industry. In this paper, a study is conducted on assessing the application potential of such clean energy power systems as solar PV, wind and fuel cells in reducing the GHG emissions of the global auto manufacturing industry. The study is conducted on the representative solar PV, wind and fuel cell clean energy systems available on the commercial market in six representative locations of GM's global facilities, including the United States, Mexico, Brazil, China, Egypt and Germany. The results demonstrate that wind power is superior to other two clean energy technologies in the economic performance of the GHG mitigation effect. Among these six selected countries, the highest GHG emission mitigation potential is in China, through wind power supply. The maximum GHG reduction could be up to 60 tons per $1,000 economic investment on wind energy supply in China. The application of wind power systems in the United States and Germany could also obtain relatively high GHG reductions of between 40–50 tons per $1,000 economic input. When compared with wind energy, the use of solar and fuel cell power systems have much less potential for GHG mitigation in the six countries selected. The range of median GHG mitigation values resulting from solar and wind power supply are almost at the same level.

KEYWORDS

greenhouse gas emissions; automotive manufacturing; clean energy supply; cost benefit analysis

1. INTRODUCTION

Automotive manufacturing is energy-intensive [1]. Greenhouse Gases (GHGs) are generated in automotive manufacturing, from both the direct on-site consumption of fossil fuel energy and indirectly from consumption of purchased electricity. As estimated, the manufacture of a typical vehicle requires approximately 120 Giga Joules of energy input [2]. In 2007, GM consumed a total of 5.543×10^{16} Joules energy in its U.S. facilities, which includes 2.149×10^{16} Joules of energy supply from purchased electricity; the total energy consumption of GM generated 6.263 million metric tons of CO_2emissions, which includes 4.359 million metric tons from purchased electricity [3].

Aware of the significance of global warming resulting from GHG emissions, the global automotive industry has worked intensively to reduce the GHG emissions from their production facilities and manufacturing processes. The Alliance of Automobile Manufacturers (AAM), formed by such major global automotive manufacturers as GM, Ford, Chrysler, Toyota, Mitsubishi, Mercedes, Porsche, Volkswagen, Volvo, and Jaguar, committed to achieve a 10% reduction in greenhouse gas emissions per number of vehicles produced from their U.S. automotive manufacturing facilities by 2012 measured from a base year of 2002 [4]. During the time period between 2002 and 2005, the AAM members has already reduced the GHG emissions intensity of their U.S. facilities, measured as CO_2 emissions per number of vehicles produced, by nearly 3% [4].

In reducing the GHG emissions, GM has implemented energy efficiency improvement efforts and conversions to lower GHG emitting fuels such as switching from coal to natural gas for the operation of boilers [3]. In statistics, GM facilities have participated in a total of 1753 improvement projects from 1991–2007, which led to a total reduction of GHG emissions of over 17 million metric tons CO_2 equivalent [3]. Through such efforts, GM has significantly improved the capacity utilization of its facilities operations. In terms of the CO_2 emission intensity of its facilities in the United States, GM has decreased its CO_2 emission per vehicle built from 2.71 metric tons/vehicle in 1990 to 2.2 metric tons/vehicle in 2007 [3]. However, despite such significant efforts in GHG mitigation, the total CO_2 emissions from GM's U.S. facilities still stand at over 6.0 million metric tons each year, due to the large volume of production [3]. Further reduction of GHG emissions from GM's production facilities is difficult since approximately 70% of the total energy consumption is from purchased electricity [3].

As the current electricity supply relies heavily on fossil fuel energy sources, further reduction of GHG emissions could be possibly achieved by using clean energy supplies to partially replace the current grid power supply of automotive production facilities, since the GHG emission intensity of clean energy technologies are much less than that of grid power supply [5]. In order to aid the global automotive industry in understanding the potential application of clean energy technologies in reducing the GHG facility emissions, a quantitative study is conducted in this project in cooperation with the General Motors Company on assessing the potential application of clean energy power systems in the efforts of GHG emission mitigation from the production facilities of global automotive manufacturers.

There are quite a few clean energy technologies for power generation available on the commercial market. Considering the specificity and requirements of automotive production facilities as well as the maturity and adaptability of the various clean energy technologies, three such clean energy power systems, namely solar PV, wind and fuel cells have good potential in stationary power supply and GHG emission reduction for automotive manufacturing industry [5]. In this study, we selected a number of representative power systems based on these three clean energy technologies to assess their potential application in GHG mitigation at automotive manufacturing facilities located at different global geographical locations.

The production facilities of global automotive manufactures are spread all over the world, with very different geographical conditions and power supply situations. For instance, GM as a leading global automotive manufacturer has its production facilities in 31 countries and has its business operations in 157 countries throughout the world [6]. In this study, we selected six representative locations from GM's major global production facility list to represent the six different regions of the world, including Detroit (United States), Mexico City (Mexico), Sao Paulo (Brazil), Shanghai (China), Cairo (Egypt) and Bochum (Germany), as shown in Figure 1 below:

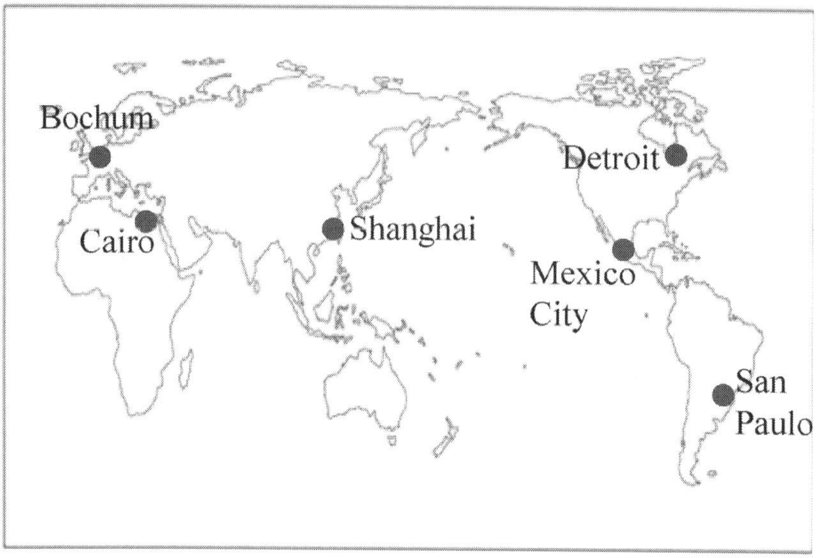

Figure 1. Six selected geographical locations of GM production facilities.

In this study, the cost benefit of using the three clean energy technologies, namely solar PV, wind, and fuel cells, was quantitatively assessed for their application in reducing the GHG emissions of automotive manufacturing facilities. The cost benefit analysis results are intended for decision support of global automotive manufacturers in selecting the appropriate clean energy technologies and optimizing the GHG mitigation effect of the same economic investment. The three clean energy technologies are benchmarked on their technology aspects, application potential (in terms of economic costs of GHG reduction) as well as the possible reduction range of GHG emissions at the six selected locations. The analysis results presented in this paper should be useful in providing detailed quantitative information, integrating the technology characteristics of clean energy power systems and geographical differences of local power generation and supply conditions, for robust decision support in GHG emission mitigation by the global automotive and similar manufacturing industries.

2. CLEAN ENERGY POWER SYSTEMS

Solar PV, wind and fuel cells have seen rapid development and deployment in the past two decades. Statistical data shows that by the end of 2008, the worldwide installed wind power capacity had reached 121 GW, and the grid-connected solar PV capacity was 13 GW [7]. Between 2004 and 2008, the global annual production of solar PV increased nearly six-fold [7]. Applications of clean energy power systems are evolving in a wide range of industrial sectors and different geographical locations around the world. Solar PV, wind and fuel cells are considered as clean energies because these power systems produce fewer emissions than the conventional grid power supply industry. However, such clean energy technologies are not completely clean. Certain amounts of emissions are still produced in various phases of their life cycle, including raw material acquisition, manufacturing, end-of-life, *etc*. When it comes to employing such clean energy technologies to replace the conventional grid power supply for the global manufacturing industry, the lifecycle GHG emissions of such clean energy technologies must be considered in the assessment of the overall GHG mitigation potential of such applications. In the section below, we will briefly introduce the solar PV, wind and fuel cell power systems selected for this study and the associated life cycle emissions for power generation and supply used in this assessment.

2.1. Solar PV System

Solar photovoltaic systems use the photoelectric effect of semiconductor materials to convert sunlight directly into electricity. The major component of PV systems is the solar module, normally a number of cells connected in series. Solar PV power systems produce negligible emissions during their operation and maintenance, but there are still emissions associated with other lifecycle phases of a solar PV power system, including raw material acquisition and production, system manufacturing, and end-of-life.

The electricity generation of solar PV depends on the solar insolation level the PV system is exposed to, which is closely linked to the geographical location where the PV system is deployed. In general, the actual power output of a Solar PV system in terms of AC electricity supply can be calculated through the following expression:

$$AC_{out} = n_m \times I_{ave} \times A_m \times e_m \times f_{\text{DC-AC}} \tag{1}$$

where:

- AC_{out}: actual power output, AC electricity (kWh)
- n_m: number of PV modules
- I_{ave}: average annual solar insolation (kWh/m^2/year)
- A_m: surface area of one PV module (m^2)
- e_m: module efficiency
- $f_{DC\text{-}AC}$: DC-AC conversion efficiency

There are quite a number of commercial solar PV systems developed by various manufacturers. These solar PV modules have different technical power generation parameters, and accordingly selection of PV modules will have a direct influence on the GHG mitigation potential results. In this study we assume there is no preference for using a specific PV module for power supply in the global automotive manufacturing. We selected five multi-crystalline solar PV modules that are considered as representative PV modules in power plant development. These five PV modules selected are the top five modules in terms of their current production volume and installed capacity in the world [8]. The detailed product information and the technical specifications of the selected five PV modules are listed in Table 1 below. In the quantitative analysis, the technical parameters of the solar PV power supplies are taken as those average values of the technical specifications of the selected five solar PV modules.

Table 1. Selected solar PV module for power supply in global automotive manufacturing.

Solar PV Module	Manufacturer	Rated Module Power (W)	Module Efficiency	Module Surface Area (m^2)
Suntech STP210-18/Ud	Suntech	210	14.30%	1.47
Sharp ND-224uC1	Sharp	224	13.74%	1.63
Qcells Q.BPPARROSOE 225	Q-Cells	225	17.00%	1.67
YL 210 P-26b/1495x990	Yingli Solar	210	14.20%	1.48
Trina Solar TSM-PC05	Trina Solar	230	14.70%	1.64

2.2. Wind Energy

Wind is another form of solar energy, generated by uneven solar heating of the earth's land and sea surfaces. Among the new renewable energy sources (excluding large hydropower), wind power is the largest addition to the renewable energy capacity [7]. Statistical data shows that the wind power installations have increased significantly in recent years, with a 29% increase in 2008, and reaching a total of 121 GW, while the installed wind power capacity in 2004 was only 48 GW [7]. The rapid increase of wind power capacity in 2008 came mainly from the increased number of installations in such countries as the United States (8.4 GW added), China (6.3 GW), India (1.8 GW), and Germany (1.7 GW) [7].

Wind power output is dependent on the wind energy density of the geographical location where the wind turbine is installed. At a specific location, environmental parameters such as wind speed and air density (related to temperature, atmospheric pressure and altitude) jointly determine the wind energy density. The wind speeds vary with the height above the surface of the earth, and the wind speed values obtained from the public reports are typically for 10 or 50 m height above the ground. Wind speed at other heights can be obtained by using the following transformation [9]:

$$v_z = v_0 \left(\frac{z}{z_0}\right)^k \qquad (2)$$

where:

- v_z: wind speed at z m height above the ground (m/s)
- v_0: wind speed at specified height of z_0 (m/s)
- z_0: specified height (m)
- k: Hellman exponent ($k = 0.34$)

In Equation (2), the Hellman exponent value is a key parameter. The k value depends on the location and the shape of the terrain on the ground and the stability of the air [10]. In urban areas, the $k = 0.34$ value is usually selected for the condition of neutral air above human inhibited areas [10]. As a result, the wind power density can be calculated through the following expression [11]:

$$W = \frac{P}{A} = \frac{1}{2} \rho \times v_z^3 \times \Gamma\left(\frac{\lambda + 3}{\lambda}\right) \qquad (3)$$

where:

- W: wind power density (W/m^2)
- P: air pressure (Pa or N/m^2)
- A: area (m^2)
- ρ: air density (kg/m^3)
- v_z: wind speed at z m height (m/s)
- λ: the dimensionless Weibull shape parameter

On the commercial market, there are quite a large number of wind turbines developed with different power capacity levels for use under different wind energy density conditions. Typical power capacity of a wind turbine for industrial scale application ranges from 1.0 MW to 4.0 MW. Different wind turbines operate differently under different wind speeds and have different power outputs corresponding to the wind speed input. As the analysis conducted in this project is to partially supply the energy needs of automotive manufacturing facilities, the wind turbines are selected on the lower end of the rated power range. In this analysis, four wind turbines of 1.5 MW rated power output are selected for analysis of their potential application in reducing the GHG emissions of the global automotive manufacturing industry. These four wind turbines are considered representative of the wind turbines at their level of capacity based on the fact that these four wind turbines have the largest installed capacity throughout the world [12]. Although wind technology is considered more environmentally benign than the conventional grid power supply, there are still environmental emissions associated with the life cycle of a wind turbine. The detailed product information and the technical specifications of the four wind turbines are shown inTable 2 below. In the analysis, we use the average values of the technical specifications of these four wind turbines as representative parameters of wind power system at the 1.5 MW level.

Table 2. Selected wind turbines for power supply in global automotive manufacturing.

Wind Turbine	Manufacturer	Rated Capacity (kW)	Rotor Diameter (m)	Sweep Area (m^2)
1.5XLE	GE	1500	82.5	5346.00
SL1500/77	Sinovel	1500	77.4	4705.13
S82 1.5 MW	Suzlon	1500	82	5281.02
S77 1.5 MW	Nordex	1500	82	5281.02

2.3. Fuel Cell

Fuel cells are electrochemical devices. A fuel cell is very much similar to a battery in its mode of electricity generation. The major difference between a fuel cell and a battery is that a fuel cell needs a continuous supply of fuels to produce electricity, while a battery has chemicals stored inside that react and produce electricity.

Fuel cells can be used for large-scale industrial power supply when sufficient amounts of fuels are supplied continuously into the system. Typical fuel cells are developed to use hydrogen (or hydrogen fuels) to produce electricity. Taking hydrogen as an example, the chemical reaction within the fuel cell system can be described as follows:

Anode side:

$$2H_2 \rightarrow 4H^+ + 4e^-$$

Cathode side:

$$O_2 + 4H^+ + 4e^- \rightarrow 2H_2O$$

Net reaction:

$$2H_2 + O_2 \rightarrow 2H_2O$$

When using hydrogen as fuel, a fuel cell power system only produces water emissions during operation, and no GHG emissions. Accordingly, the use of fuel cell power systems can significantly reduce the GHGs emissions compared to conventional power supplies. However, currently there are quite a number of challenges associated with hydrogen fuel production, storage and use in fuel cells. As a result, natural gas is more commonly used in stationary fuel cell power systems. Although the natural gas consumed in a fuel cell power system is only involved in the oxidation process, instead of burning, the carbon elements of natural gas are still converted into CO_2 emissions from the following reaction process:

$$CH_4 + 2H_2O = CO_2 + 4H_2$$

If a natural-gas based fuel cell power system will be used, a quantitative trade-off analysis must be conducted to evaluate the GHG mitigation potential of the fuel cell power system, between the life cycle GHG emissions of a fuel cell power system and the GHG emissions of the conventional grid power supply, based on the same amount of electricity consumption.

Fuel cells have an advantage over solar and wind energy in that fuel cells are not limited by geographical conditions. The power efficiency of the fuel cell power systems can always be maintained at the rated output level and it can provide a steady power supply as long as sufficient fuels are supplied continuously into the system. Currently, fuel cells are mainly developed as mobile energy sources for transportation applications. For stationary power generations, there are only a few models available in the United States. In this analysis, two fuel cell stationary power systems are selected, with one system using hydrogen fuel, and the one using natural gas. The selected fuel cell power systems are listed in Table 3 below, with their detailed product information and technical specifications.

Table 3. Selected fuel cell stationary power system for electricity generation.

Stationary Fuel Cells	Manufacturer	Fuel Type	Rated Power (kW)
PureCell 200	UTC Power	Natural gas	200
Nedstack PS100	Nedstack	Hydrogen	100

3. GHG EMISSION MITIGATION THROUGH CLEAN ENERGY SUPPLY

The selected three clean energy technologies have different GHG mitigation potential at different locations. Solar PV and wind power systems are dependent on geographical conditions in particular. During the assessment of the application potential of the three clean power systems, the geographical conditions of these six locations are integrated into the analysis to calculate the actual power output of solar PV and wind power systems. The application potential of the three clean power systems are analyzed in three aspects: capacity factor of the clean energy power system, amount of GHG reduction on a unit economic cost scale, and the potential reduction range of each clean energy system in the six countries selected.

The geographical differences of solar energy among the six selected cities are characterized by the average amount of the total solar radiation incident on a horizontal surface on an annual base. The solar insolation data used in this analysis were the data collected by NASA for that geographical location during a 22-year time period (from July 1983 to June 2005) [13]. The solar PV module efficiency used in the analysis is the average of the five efficiency values of the selected solar PV modules in Table 1, namely 14.79%, as a representative of solar PV in that class. Since the solar PV power system produces DC power output, for supplying automotive manufacturing the solar PV DC power output must be converted to AC power prior to its actual use. In this analysis, the conversion efficiency from DC to AC power supply is taken as the typical value of 77% [14].

The geographical differences of wind energy are characterized by the differences of wind speed and wind power density among the six selected locations. The wind speed data for the six selected cities are the data collected by NASA during a 10-year time period (from July 1983 to June 1993) [15]. The NASA wind speed data are collected only for 50 m height above the ground, while the average height of the selected four wind turbines is 80 m. In our analysis, the NASA 50 m data are transformed to 80 m wind speed data using Equation (2):

The solar and wind power density values for the six selected cities are shown in Table 4 below. For solar energy application, the results show that Cairo has the highest potential while Bochum has the lowest; for wind energy, Bochum has the highest potential while Mexico City has the lowest. Since fuel cell power systems depend on fuel supply instead of geographical conditions, they are not listed here.

Table 4. Solar and wind power density in the selected six locations [9].

	Detroit	Mexico city	Sao Paulo	Shanghai	Cairo	Bochum
Solar insolation ($kw/m^2/year$)	1269.90	1890.70	1661.36	1410.73	1929.03	986.11
Wind power density (w/m^2)	356.36	76.50	116.59	352.08	213.01	591.78

3.1. Capacity Factor of Solar and Wind Power Systems

Although the energy density information can serve as the basic indicator of the application potential of clean energy power systems, further analyses are needed to understand more about the actual implementation of these clean energy power systems in different geographical locations. For solar and wind energy, we employ the capacity factor, a meaningful metric to assess their actual application potential in a specific geographic location, to assess the performance the clean energy systems at the selected locations. The capacity factor is defined as the ratio between the actual power output and the total rated power of the system available for the power generations. The expression of capacity factor is shown in Equation (4) below:

$$CF = \frac{P_{output}}{P_{system}} \times 100\% \tag{4}$$

where:

- CF: capacity factor of a power supply system
- P_{output}: actual power output of the supply system (kW)
- P_{system}: rated power of the supply system (kW)

The calculated capacity factors for solar PV and wind power systems at the six selected geographical locations are shown in Figure 2. The actual power output for solar PV system is calculated by multiplying the local annual solar insolation, the module area, module numbers, module efficiency and the DC to AC conversion efficiency. The actual power output for wind turbines is calculated by applying the power curve of the wind turbine model. The power curve of a specific model of wind turbine is fitted by the power output values at different wind speeds supplied by the wind turbine manufacturer. The results demonstrate that for current solar PV power systems, the actual capacity factors are all below 20% at these six selected locations; while the capacity factor of wind can reach up to 47% at the six selected locations, depending on the local wind energy resources.

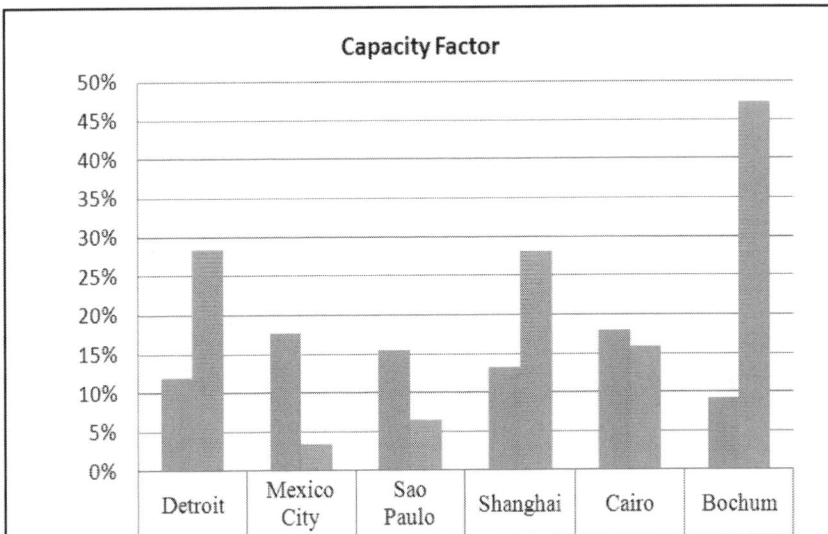

	Detroit	Mexico City	Sao Paulo	Shanghai	Cairo	Bochum
Solar PV	11.84%	17.62%	15.48%	13.15%	17.98%	9.19%
Wind	28.42%	3.39%	6.48%	28.09%	15.97%	47.24%

Figure 2. Capacity factor of solar PV and wind power systems at the six selected locations.

From Figure 2, we can see that Cairo has the largest capacity factor for solar PV at 17.98%, followed by Mexico City (17.62%) and Sao Paulo (15.48%). The capacity factors for solar PV in Shanghai, Detroit and Bochum are 13.15%, 11.84% and 9.19%, respectively. From the comparison of the capacity factors at the six locations, it can be concluded that Cairo has the largest technical potential, followed by Mexico City, Sao Paulo, Shanghai, Detroit and Bochum.

For wind power systems, we find that Bochum has the largest wind capacity factor at 47.24%, followed by Shanghai (28.09%), Detroit (28.42%), Cairo (15.97%), Sao Paulo (6.48%) and Mexico City (3.39%). For the Equation (5) used in capacity factor calculations, we use the rated power as the denominator, which means for the same rated power, the larger the output power, the larger the capacity factor. For a wind turbine, the input power can be simply estimated by the following equation:

$$P_i = D_w \times A_S \tag{5}$$

where:

- P_i: input power of a wind turbine (W)
- D_w: the wind power density (W/m^2)
- A_S: the sweeping area of turbine blades (m^2)

From Equation (5), it can be seen that the larger the wind power density, the larger the input power. Obviously, for two wind turbines with the same efficiency, the output power is positively correlated with the input power under the wind speed conditions before the rated power is reached. From this point of view, the geographical locations with a high wind power density can lead to a large power output of a wind turbine and accordingly produce a large capacity factor. In the previous energy density analysis as indicated in Table 4, the results demonstrate that Bochum has the highest wind power density, followed by Shanghai, Detroit, Cairo, Sao Paulo and Mexico City. The capacity factor analysis results as shown inFigure 2 are consistent with the energy density results shown in Table 4.

3.2. Economic Analysis of GHG Mitigation

To promote the application of clean energy supply in industrial operations for GHG mitigation, the economic performance of the mitigation efforts must be understood and quantitatively assessed to assist decision-making in industrial sustainability management and practices. In this section, we provide quantitative assessment of the GHG mitigations through the three clean energy supply patterns at the six selected geographical locations, in terms of the amount of GHGs to be reduced on a base scale of $1,000 economic input (tons GHG reduction/$1,000).

Currently, the economic costs of solar PV, wind and fuel cells are very different, but the economic costs of clean power systems within the same category, such as the five selected solar PV modules, are approximately on the same level due to the international competition between their manufacturers. In this analysis, the clean energy power systems are selected as average of the representative power systems of each clean energy type available on the commercial market. For the economic costs of each type of clean power system, we use the average economic costs data statistically collected and recently released by U.S. Energy Information Administration (EIA) for estimating the economic performance of GHG mitigation efforts through clean energy supply [16]. The economic cost data used on the solar PV, wind, and fuel cell power systems include overnight cost of each clean power system and the associated variable and fixed O&M costs [16]. The economic costs and the LCA GHG emissions used on the three clean energy power systems are shown in Table 5 below.

Table 5. Costs and GHG emissions of clean energy power systems.

	Overnight Cost ($/kW)	Variable O&M ($/kWh)	Fixed O&M ($/kW)	LCA GHG Emissions (g/kWh)
Solar PV	6171	0.00	11.94	72.4 [1]
Wind	1966	0.00	30.98	10.84 [2]
Fuel Cell (NG)	5478	0.049	5.78	683 [3]
Fuel Cell (H_2)	10,735 [5]	0.00 [5]	2147 [5]	83 [4]

[1] Multicrystalline Si Solar Cell, LCA result from reference [17]; [2] 1.5 MW onshore wind turbine, LCA result from reference [18]; [3] Stationary Fuel Cell (natural gas), LCA result from reference [19]; [4] Stationary Fuel Cell (hydrogen), LCA result from reference [20]; [5] The cost data for hydrogen PEM fuel cells, from reference [21].

As a result, the cost benefit of the GHG mitigation through clean energy supply is assessed by using the following expression:

$$G = \frac{(E_{local} - E_i) \times A_i \times T_i}{(C_{Ni} + C_{Fi}) \times A_i + C_{Vi} \times (T_i \times A_i)} \qquad (6)$$

where:

- G: the amount of GHG reduction (ton/$1,000)
- E_{local}: emission factor of GHGs from local grid power supply (kg/kWh)
- E_i: the life cycle GHG emissions of clean energy i (kg/kWh)
- A_i: total installed capacity of clean energy power system i
- T_i: operational life time of clean power system i (h)
- C_{Ni}: overnight cost of clean power system i ($/kW)
- C_{vi}: variable O&M cost of clean power system i ($/kWh)
- C_{Fi}: fixed O&M cost of clean power system i ($/kWh)

The GHG emission factors, E_{local}, used in the economic analysis are the total CO_{2eq} values calculated based on the IPCC Global Warming Potential (GWP), as shown in Table 6 below for the six selected cities.

Table 6. GHG emission factors of local grid power supply.

Region	Emission Inventory (g/kWh)			
	CO_2	CH_4	N_2O	CO_{2eq} [3]
USA [1]	676	0.01815	0.01053	680
Mexico [2]	593	0.01676	0.00230	594
Brazil [2]	93	0.00251	0.00106	93
China [2]	839	0.01458	0.01841	845
Egypt [2]	436	0.01365	0.00177	437
Germany [2]	539	0.00637	0.00779	542

[1] The U.S. electricity emission factors are from reference [22]. [2] The international electricity emission factors are from reference [23]. [3] The CO_{2eq} are calculated based on the GWP values from reference [24].

In the economic analysis, the amounts of GHG reductions are calculated separately for natural-gas fuel cells, and hydrogen fuel cells. The results (tons GHG reduction/$1,000 economic input) are shown in Figure 3 below. The calculated results in Figure 3 show the different mitigation effects of clean energy supply patterns based on the same amount of investment scenario at the six selected global locations.

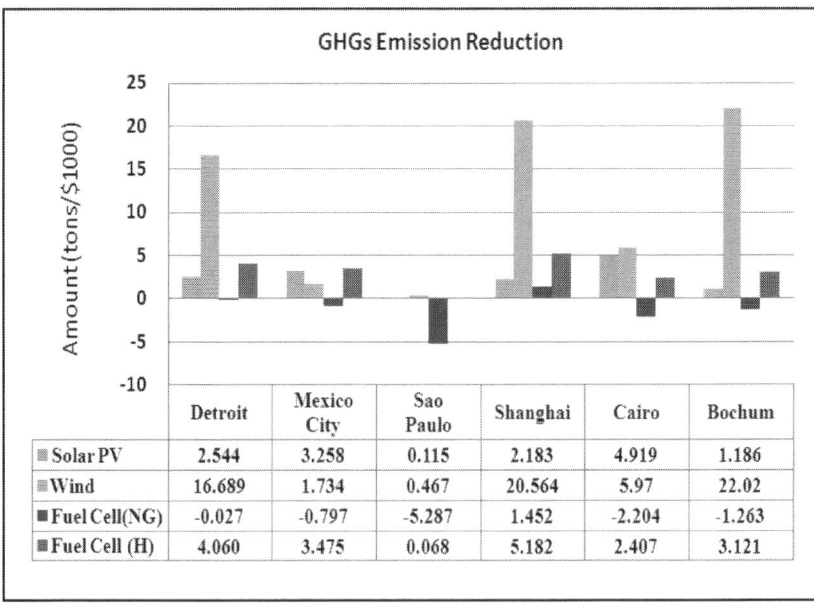

Figure 3. Economic analysis of GHG mitigation through clean energy supply.

The calculated results in Figure 3 indicate that wind has the greatest application potential among the three types of clean energy power systems for GHG emission mitigation, in particular at those locations with high wind energy density and/or high GHG emissions from local grid power supply. Application of wind in Bochum and Shanghai can achieve a GHG reduction over 20 tons per $1,000 economic input, while the application of wind in such cities as Mexico City and Sao Paulo has almost negligible mitigation effects.

Although solar PV systems are considered very clean in the usage phase, the high cost of PV systems and the relatively high emissions from their manufacturing and material production stages make solar PV much less preferable than wind for GHG emission mitigation. From Figure 3, the best mitigation effects from solar PV systems are only at the level of a quarter of the mitigation effects from wind applications, based on the same economic input scale.

The economic performance of fuel cell power systems is closely related to the type of fuel used. The natural gas based fuel cells, due to the CO_2 emission from the consumed natural gas, in most cases will further increase the total GHG emissions. The results in Figure 3 indicate that the GHG mitigation through natural gas based fuel cell power system is only feasible in Shanghai, China, because of the high GHG emission factor of local power supply industry [23]. In all other five locations, using natural gas based fuel cell power systems will increase the amount of GHG emissions.

The hydrogen fuel cell power system, based on the calculated results in Figure 3, can achieve a mitigation effect on GHG emissions between solar and wind at the six selected locations. This is partially due to the high cost of the hydrogen based fuel cell power systems and the high energy density required for hydrogen production. It is expected that in the future the hydrogen powered fuel cells would have a greater application potential as the economic costs of the systems are lowered further and after better ways for hydrogen production and storage are identified.

As there are technical variations among the selected clean energy power systems, the cost benefit of using a specific clean energy technology might be different from each other. In order to characterize the sensitivity and uncertainty of using a specific power system among the selected models for each clean energy technology, the range of GHG reduction, in the unit of tons/$1,000, for the selected clean energy power systems are calculated for their applications at the six selected locations. The results are shown in Figure 4 below, with the median value indicated for each mitigation range.

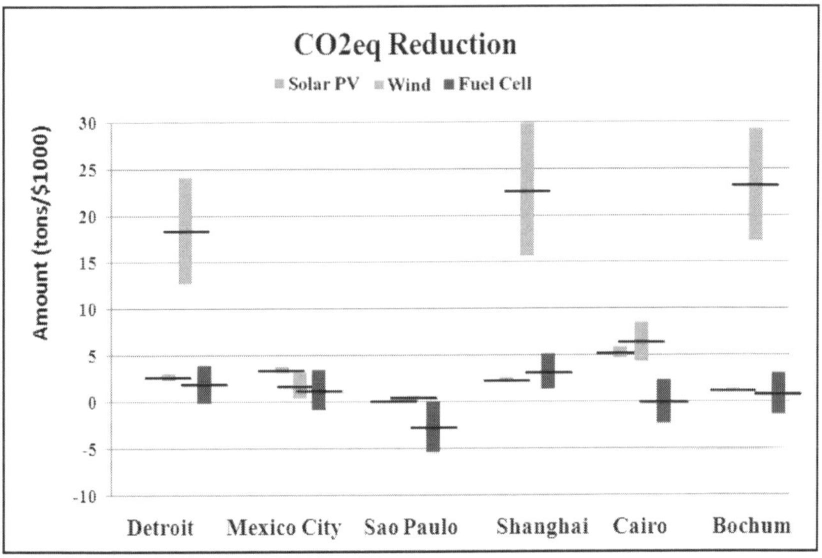

Figure 4. Range of GHG mitigation potential through clean energy supply at the selected six locations.

Figure 4 demonstrates that the selected wind power systems have the highest GHG mitigation potential among the three clean energy technologies, in particular for application in Detroit, Shanghai and Bochum. The highest GHG reduction from wind application in Shanghai can reach 30 tons per $1,000 economic input. Wind power supply in Bochum and Detroit can achieve up to 29 tons and 24 tons of reduction, respectively, per $1,000 economic input. As quantitatively indicated, the minimum amounts of GHG reduction from adoption of the selected wind turbines in Detroit, Shanghai and Bochum are still more than 13, 15, and 17 tons, respectively, per $1,000 economic input. When compared, the application of the selected solar PV models can only reduce an amount of GHG emissions less than 6 tons per $1,000 economic input. Fuel cell power systems are almost at the same level of GHG mitigation effect with solar PV systems. If using natural gas based fuel cell power systems, GHG mitigation is only feasible in Shanghai, China.

3.3. Range of GHG Mitigation Potential at Different Regions

The above analysis results are on applications of the selected representative clean power systems at specific geographical locations (cities). In order to understand the GHG mitigation potential of clean energy supply for broader areas, we have extended the analysis results to the country-wide geographical area, and assessed the range of GHG mitigation potential in the six countries selected for this study. The range of GHG mitigation potential are calculated for solar PV and wind power systems based on the selected average technical parameters shown in Table 1 and Table 2 above. The range of GHG mitigations from solar PV power supply is calculated by considering the best and worst power generation scenarios under the highest and lowest solar insolation conditions within the geographical boundary of the selected country. The range of GHG mitigation from wind power supply is calculated by considering the best and worst power generation scenarios under the highest wind energy density of that country and the minimum wind speed (4.47 m/s) required for wind turbine installation [25]. The range of GHG mitigation from fuel cell power supply is calculated by considering the fuel differences of the power systems based on the technical parameters shown in Table 3. The calculated range of GHG mitigation potential gives the maximum and minimum amount of GHGs which can be mitigated through each clean energy supply pattern in these six countries on the basis of the same economic input. The results are shown in Figure 5 below, with the median value indicated for each mitigation range.

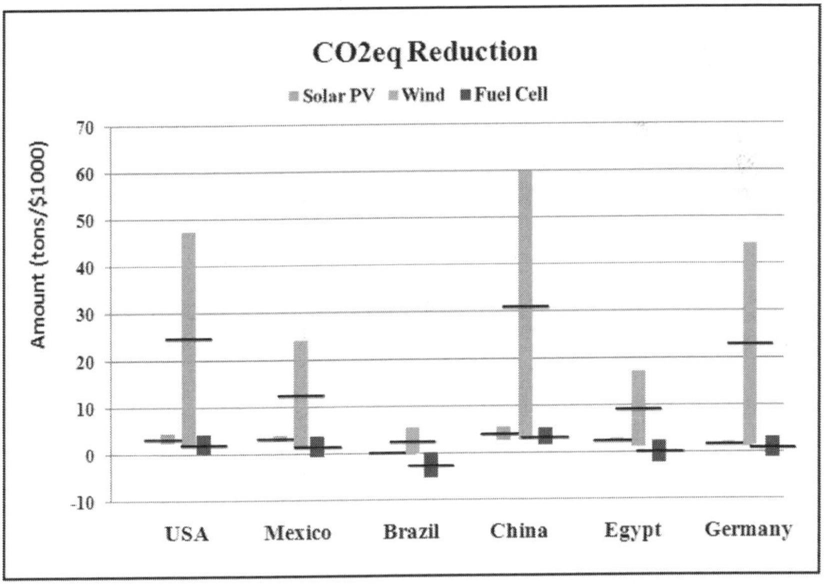

Figure 5. Range of GHG mitigation potential through clean energy supply in the selected six countries.

The calculated results in Figure 5 demonstrate that the best GHG mitigation opportunity is in China. With $1,000 economic investment, the maximum amount of GHG reduction can be as high as 60 tons, while application of wind power systems in the United States and Germany can also obtain a maximum GHG reduction of between 40 and 50 tons. When compared with the wind supply pattern, application of solar and fuel cell power systems has much less potential for GHG mitigation in each country. The median values of GHG mitigation range from fuel cells and wind power supply are almost at the same level.

The maximum reduction of GHG emissions through clean energy supply depends on many factors. From the results of our analysis, the most important factors for an optimal GHG mitigation are on the selection of the clean energy technology and the geographical location for system installation. In this analysis, the technical differences of the selected power systems in each clean energy category are not fully assessed and benchmarked, but such differences are believed having very small influences on the decision-making in clean energy technology selections.

4. CONCLUDING REMARKS

GHG emissions are a global issue due to the long persistence time of GHGs in the atmosphere. Any emissions generated at different locations in the world ultimately contribute to the same global warming problem. Clean energy systems have been recognized widely for their potential in mitigating the GHG emissions from the current grid power supply. However, due to the differences of energy supply structure and geographical conditions of different regions, application of clean energy systems in different locations can result in different mitigation effects.

Application of clean energy power systems to partially supply the energy needs of global automotive and similar manufacturing facilities can greatly reduce the significant amount of GHG emissions from these production facilities and manufacturing processes. However, there is a lack of cost benefit data and information for decision support in sustainable practices of global manufacturing industry. In this paper, we present our study on the GHG mitigation potential of supplying clean energy power systems such as solar PV, wind and fuel cells for the global automotive manufacturing industry, in cooperation with the General Motors Company.

The study is performed on the representative clean energy power systems currently available on the market. The selected geographical locations, including Detroit (United States), Mexico City (Mexico), Sao Paulo (Brazil), Shanghai (China), Cairo (Egypt) and Bochum (Germany) are major production centers for GM and are considered representatives of each geographical region in the World in this analysis. The mitigation potential of each clean energy system is quantitatively investigated with consideration of both geographical conditions of each location and technical characteristics of each clean power system.

The study of GHG mitigation potential through clean energy supply is conducted in three aspects: capacity factor of each clean energy system at the different locations, economic performance of GHG mitigation efforts, and the maximum mitigation potential in the selected country. The results demonstrate that wind power is superior to other two clean energy technologies in the economic performance of the mitigation effect, particularly in those regions with high wind energy density (such as Bochum) and/or high GHG emission factor (such as Shanghai).

Considering the overall conditions for GHG mitigation and the economic cost factor, the study shows that the highest mitigation potential of GHG emissions among these six selected countries is in China, through wind power supply. The maximum GHG reduction could be as high as 60 tons per $1,000 economic cost investment in wind energy supply in China. The application of wind power systems in the United States and Germany may also achieve a maximum of GHG reduction of between 40 and 50 tons per $1000 economic input. When compared with the wind energy supply, application of solar and fuel cell power systems has much less potential for GHG mitigation in the six countries selected for study. The median values of GHG mitigation range resulting from solar and wind power supply are almost at the same level.

ACKNOWLEDGEMENTS

The financial support from the internal funding of the University of Wisconsin, MI, General Motors Company, and the Chinese NSFC (Project No. 51075415) are greatly appreciated.

REFERENCES

1 Yuan, C.; Zhang, T.; Rangarajan, A.; Dornfeld, D.; Ziemba, B.; Whitbeck, R. A decision-based analysis of compressed air usage patterns in automotive manufacturing. *J. Manuf. Syst.* **2006**, *25*, 293–300.

2 Maclean, H.; Lave, L. A life-cycle model of an automobile. *Environ. Policy Anal.* **1998**, *3*, 322A–330A.

3 General Motors Company. *Public Policy Center Environment and Energy*, Voluntary Reporting of General Motors Corporation United States Greenhouse Gas (GHG) Emissions for Calendar Year (1990–2007). 2008. Available online: http://www.gm.com/corporate/responsibility/environment/reports/greenhouse_gas_emissions/ghgreport_2007.pdf(accessed on 5 October 2010).

4 U.S. D.O.E. (Department of Energy). *Climate Vision Progress Report*; 2007. Available online: http://www.doe.gov/media/ Climate_ Vision_ Progress_Report.pdf (accessed on 2 October 2010).

5 Yuan, C.; Dornfeld, D. Reducing the environmental footprint and economic costs of automotive manufacturing through an alternative energy supply. *Trans. North Am. Manuf. Res. Inst.* **2009**, *37*, 427–434.

6 General Motors Company. General Motors: About GM 2010. Available online: http://www.gm.com/corporate/about/(accessed on 12 October 2010).

7 REN21. *Renewables Global Status Report*, 2009. Available online: http://www.ren21.net/pdf/RE_GSR_2009_Update.pdf(accessed on 4 October 2010).

8 Renewable Energy World. *Top 10: Ten Largest Solar PV Companies*, 2010. Available online: http://www.renewableenergyworld. com/rea/blog/post/2010/06/top-10-ten-largest-solar-pv-companies (accessed on 23 October 2010).

9 Simiu, E.; Scanlan, R.H. *Wind Effects on Structures*; Wiley: New York, NY, USA, 1978; p. 47.

10 Kaltschmitt, M.; Streicher, W.; Wiese, A. *Renewable Energy: Technology, Economics, and Environment*; Springer: New York, NY, USA, 2007; p. 55.

11 Chang, T.J.; Wu, Y.T.; Hsu, H.Y.; Chu, C.R.; Liao, C.M. Assessment of wind characteristics and wind turbine characteristics in Taiwan. *Renew. Energy* **2003**, *28*, 851–871.

12 My Wind Power System. *The 10 Major Wind Power Companies in the World*, Available online: http://www.mywindpowersystem.com/2009/04/the-10-major-wind-power-companies-in-the-world/ (accessed on 23 October 2010).

13 NASA Atmospheric Science Data Center. *Surface Meteorology and Solar Energy, A Renewable Energy Resource Web Site (Release 6.0)*; 2008. Available online: http://eosweb.larc.nasa.gov/cgi-bin/sse/sse.cgi?na+ s01+s06#s01 (accessed on 12 October 2010).

14 Marion, B.; Anderberg, M.; Gray-Hann, P. *Recent Revisions to PVWATTS. NREL/CP-520-38975*; National Renewable Energy Laboratory: Golden, CO, USA, 2005.

15 NASA Atmospheric Science Data Center. *Surface Meteorology and Solar Energy, A Renewable Energy Resource Web Site (release 5 Data Set)*; 2005. Available online: http://eosweb.larc.nasa.gov/ sse/ global/ text/10yr_wspd50m (accessed on 12 October 2010).

16 U.S. E.I.A. (Energy Information Administration). *Cost and Performance Characteristics of New Central Station Electricity Generating Technologies*; 2010. Available online: http://www.eia. doe.gov/oiaf/ aeo/excel/aeo2010%20tab8%202.xls(accessed on 12 September 2010).

17 Azzopardi, B.; Mutale, J. Life cycle analysis for future photovoltaic systems using hybrid solar cells. *Renew. Sustain. Energy Rev.* **2010**, *14*, 1130–1134.

18 Pehnt, M. Dynamic life cycle assessment (LCA) of renewable energy technologies. *Renew. Energy* **2006**, *31*, 55–71.

19 Jaap, V.; Rooijen, A. Life Cycle Assessment of the PureCellTM Stationary Fuel Cell System: Providing a Guide for Environmental Improvement. M.Sc. Thesis, Delft University of Technology, Delft, The Netherlands, 2006.

20 Viebahn, P.; Krewitt, W. *Environmental and Ecological Life Cycle Inventories for Present and Future Power Systems in Europe (ECLIPSE)*; European Commission, DG Research (Project No. NNE5-2001-00071): Brussels, Belgium, 2003.

21 Gerboni, R.; Pehnt, M.; Viebahn, P.; Lavagno, E. *Final Report on Technical Data, Costs and Life Cycle Inventories of Fuel Cells. New Energy Externalities Developments for Sustainability (NEEDS)*; NEEDS Project: Rome, Italy, 2008. Available online: http://www.needs-project.org/RS1a/RS1a%20D9.2%20Final%20report%20on%20fuel%20cells.pdf (accessed on 15 September 2011).

22 U.S. E.I.A. (Energy Information Administration). *Domestic Electricity Emission Factors,1999–2002*; EIA: Washington, DC, USA, 2007. Available online: http://www.eia.doe.gov/ oiaf/1605/excel/electricity_ factors_99-02region.xls (accessed on 10 September 2010).

23 U.S. E.I.A. (Energy Information Administration). *ternational Electricity Emission Factors by Country,1999–2002*; EIA: Washington, DC, USA, 2007. Available online: http://www.eia.doe.gov/ oiaf/1605/excel/ electricity_factors_99-02country.xls (accessed on 10 September 2010).

24 IPCC. *IPCC Fourth Assessment Report (AR4) by Working Group 1 (WG1)*; IPCC: Geneva, Switzerland, 2007. Available online: http://www.ipcc.ch/pdf/assessment-report/ar4/wg1/ar4-wg1-chapter2.pdf (accessed on 10 September 2010).

25 American Wind Energy Association (AWEA). 2010. Available online: http://www.awea.org/faq/rsdntqa.html(accessed on 13 September 2010).

CHAPTER 3

Governing China's Clean Energy Transition: Policy Reforms, Flexible Implementation and the Need for Empirical Investigation

Kevin Lo

Department of Geography, Hong Kong Baptist University, Kowloon Tong, Hong Kong, China; Tel.: +852-3411-3397; Fax: +852-3411-5990

ABSTRACT

In the ten years since committing to clean energy transition, China has formulated a large number of policies and programs to achieve some very ambitious targets. This paper argues that the dearth of empirical studies concerning the implementation of these new policies and programs has created a knowledge gap between official policy documents, which are vague and lacking in specifics, and official policy outcomes, which are unreliable. In particular, the merits and limitations of flexible implementation with regard to desirable outcomes need to be debated and clarified. This paper calls for more empirical investigation in four areas as a starting point: (1) the nature and extent of flexibility in the implementation; (2) implementation strategies and their impacts; (3) factors that shape the behavior of local officials responsible for implementation; and (4) the relationship between the central-local relation and policy implementation.

KEYWORDS

clean energy transition; policy; flexible implementation; China

1. INTRODUCTION

Traditionally, the energy system of China has been characterized by a heavy reliance on coal and high energy intensity. Coping with the associated economic and environmental problems, of both a domestic and an international nature, has become a top priority for the government. The 11th Five-Year Plan (2006–2010), in which the goal of reducing national energy intensity by 20% in five years was announced, marked the beginning of China's clean energy transition. Various other clean energy objectives have since been issued, the latest, to be achieved by 2020, being documented in the Energy Development Strategic Plan (2014–2020). These objectives include limiting energy consumption to 4.8 billion tons of standard coal equivalent (tce), limiting coal consumption to 4.2 billion tons, increasing the proportion of natural gas to 10%, increasing non-fossil-fuel energy to 15% and reducing the coal proportion to less than 62% [1]. In the longer term, China has pledged in the United Nations Framework Convention on Climate Change (UNFCCC) to achieve peak carbon emissions by approximately 2030 [2].

Whether and how China will be able to meet these ambitious goals depends on the effective governance of clean energy transition. There is no doubt that, in a relatively short time, the Chinese government has radically changed its priorities from inexpensive energy to clean and renewable energy and from accommodating the growth of energy consumption to constraining energy consumption. This shift in priorities was accompanied by a large number of new clean energy policies and programs. Given the rapidly changing policy context, this paper argues that the lack of empirical research concerning the implementation process has limited our understanding of China's clean energy transition. Consequently, a knowledge gap has been created between official policy documents, which are often vague and lacking in specifics, and official policy outcomes, which are unreliable, if not outright impossible. In particular, the recent emphasis on flexible implementation, in the sense that local governments are allowed, and even encouraged, to adapt central directives to suit local conditions, has increased the importance of implementation *vis-à-vis* policymaking. Consequently, the merits and limitations of flexibility with regards to desirable environmental outcomes need to be debated and clarified.

A better understanding of China's clean energy transition is of international interest for several reasons. First, China is now the world's largest annual energy consumer and greenhouse gas emitter [3]; therefore, the activities of the country have global implications in terms of climate change, energy security, energy trade and the geopolitics of energy. Second, as the world's largest developing country, China's experience can serve as a model for fellow developing countries about to embark on a similar process of clean energy transition. Third, as an emerging world power, China wields significant influence on global climate negotiations; this was apparent during the 2009 Copenhagen Climate Summit [4,5]. Countries such as the United States and Australia have used the perceived inaction of China and other developing countries as justification to disregard their own mitigation responsibilities. Consequently, the commitment to clean energy transition demonstrated by China can provide much-needed momentum to the efforts for attaining a constructive and cooperative global solution to climate change.

The remainder of this paper will elaborate on the argument by first describing China's clean energy transition policy from two aspects: energy efficiency and conservation, to reduce energy use through technological and behavioral changes; and renewable and low-carbon energy, to move from fossil fuels to renewable energy (e.g., wind, solar, hydro and geothermal) and low-carbon energy (e.g., nuclear and natural gas). Then, it will examine empirical studies on the implementation process, noting both their strengths and limitations, identify the key debate and suggest certain important research areas for future studies.

2. ENERGY EFFICIENCY AND CONSERVATION POLICY

China's intention with regard to energy efficiency and conservation was announced by the State Council with its mandatory national target of a 20% reduction in energy intensity between 2006 and 2010, as well as the revision of the Energy Conservation Law in 2007. To achieve these objectives, the central government has developed a number of new clean energy programs that target the industrial, building and transport sectors.

The most important program supporting industrial energy efficiency and conservation is the Ten Thousand Enterprise Low-Carbon Energy Conservation Programme (Ten Thousand Enterprises Programme), introduced in 2011. This command-and-control program regulates more than 16,000 energy-intensive enterprises that collectively account for 60% of the country's total energy consumption [6]. It requires these enterprises to achieve certain amounts of energy saving and to implement a number of energy management measures, for which funding of 240–300 CNY/tce (1 CNY = 0.15 USD) is provided. The program is thus expected to deliver 250 million tce of energy savings over five years, more than one-third of China's total energy-saving target of 670 million tce.

In addition to the Ten Thousand Enterprises Programme, the central government has prioritized the elimination of a backward-production capacity, that is production equipment and processes that are deemed overly energy intensive and polluting. This policy measure consists of both economic and administrative elements. Under the first, the central government imposes an electricity surcharge of 0.05–0.4 CNY/kWh on industries that fall within the elimination category [7,8], and local governments are empowered to impose a higher surcharge: for example, in 2015, Beijing increased this surcharge to 0.2–0.5 CNY/kWh [9]. The second mechanism is forced closure [10], which shuts down a certain amount of old production capacity. The central government does offer compensation to affected enterprises, however, due to the negative impact on the workers. Between 2006 and 2010, the central government paid 21.91 billion CNY in compensation, successfully closing down production capacity in relation to 122 million tons of iron, 70 million tons of steel and 330 million tons of cement [11].

Turning to the building sector, the central government initiated a subsidy-based program in 2007 to support the retrofitting of insulation in existing buildings: 55 CNY/m^2 in severely cold regions and 45 CNY/m^2 in cold regions. This funding can be used for improving the thermal properties of the building envelope (*i.e.*, wall, roof and windows), installing heat meters and improving the efficiency of heat distribution systems. To date, the program has enabled the successful retrofitting of 182 million m^2 between 2007 and 2010 and 750 million m^2 between 2011 and 2014 [12], making it one of the most successful building-retrofit programs in the world. New buildings in China have to comply with the energy building codes, which set minimum energy-efficiency requirements, and are quite stringent compared to international standards [13]. Enforcement of the energy building codes has also significantly improved recently, with an official compliance rate of 100% being achieved for the first time in 2010 [14].

With regard to the transport sector, the central government has shown great enthusiasm for promoting electric vehicles, which are regarded as the future [15]. Since 2009, the central government has been offering subsidies to purchase plug-in hybrid and purely electric vehicles. According to the latest update, the subsidy offered is 23,000–55,000 CNY per vehicle, depending on the type; however, this will be reduced by 20% in 2017/2018 and 40% in 2019/2020 [16]. An added incentive is that the owners of eligible electric vehicles are exempt from the annual Vehicle and Vessel Tax [17]. However, the effect of these programs has been less than expected, and at the end of 2014, there were only 120,000 new energy vehicles, lagging far behind the 2015 target of 500,000 [18]. This low rate of uptake reflects concerns related to the convenience of charging points, distance limitations, battery longevity and the high price of the vehicles [19]. In addition, tightening of the Fuel Economy Standards has been playing an important role in rendering combustion engine vehicles, which numbered 150 million at the end of 2014, more energy efficient. The Fuel Economy Standards were introduced in 2005 and have since been revised three times. The latest edition, GB 19578-2014 [20], which will be in effect from 2016 to 2020, was designed to improve the fuel economy of new passenger vehicles to 5.0 L/100 km by 2020. Moreover, generous Vehicle and Vessel Tax incentives are being offered for vehicles that are more efficient than the standards require [17].

3. RENEWABLE AND LOW-CARBON ENERGY POLICY

Since the Renewable Energy Law was passed, followed by the Mid- to Long-Term Development Plan for Renewable Energy in 2007 [21], renewable energy has become a priority for energy development in China. Similar to many other countries worldwide, the rapid development of renewable energy in China is driven mainly by feed-in tariffs. Initially, the level of these tariffs was determined through auctions to encourage competition among the developers. However, this resulted in bidding wars, which was detrimental to the long-term sustainability of the renewable energy industry. Driven by aggressive pricing from state-owned enterprises, the winning bids for wind power during this period ranged from 0.373 CNY/kWh to 0.519 CNY/kWh, which were clearly unprofitable for the suppliers [22]. In response, a standard feed-in tariff system, ranging from 0.51 CNY/kWh to 0.61 CNY/kWh (adjusted to 0.49–0.61 CNY/kWh in 2015), was introduced in 2009 to replace the auctions for onshore wind generation [23]. This new system has successfully diversified the investment in wind power: by the end of 2013, more than 1300 companies had invested in wind power, with state-owned enterprises accounting for 81% of the installed capacity, local private enterprises 3.8%, overseas private enterprises 2.6% and joint ventures 12.7% [24].

The success of the wind-power feed-in tariff system has encouraged the government to use this policy to support other forms of renewable energy, including: bioenergy in the agricultural sector and forestry in 2010 (0.75 CNY/kWh) [25], utility-scale photovoltaics in 2013 (0.90–1.00 CNY/kWh) and offshore wind power in 2014 (0.75–0.85 CNY/kWh) [26]. Currently, distributed photovoltaics also receive a fixed-rate subsidy of 0.42 CNY/kWh. It should also be noted that several local governments have established subsidies that complement the national feed-in tariff in an effort to attract investment in renewable energy: for example, the Shanghai Municipal Government offers a five-year support program for onshore wind-energy generation (0.1 CNY/kWh), offshore wind-energy generation (0.2

CNY/kWh) and photovoltaics (0.25–0.4 CNY/kWh) [27]. The feed-in tariff system in China is financed primarily by the renewable energy surcharge, which is levied nationally, except in the Tibetan Autonomous Region, on all types of electricity consumption. When firstly introduced in 2006, the surcharge was relatively modest at a rate of 0.001 CNY/kWh; however, the rate has since been increased several times, as the collected funds were insufficient to cover the feed-in tariff expenditure. Currently, the levy is 0.015 CNY/kWh, except for residential and agricultural uses, which remain at 0.008 CNY/kWh [28].

In addition to the feed-in tariff, renewable energy is further supported by tax incentives. The hydroelectricity sector, which does not benefit from feed-in tariffs, is eligible for a value-added tax rebate: 8% for utility-scale installations and 3% for small-scale installations, which are substantially lower than the normal 17% rate [29,30]. The wind and solar sectors benefit from a 50% value-added tax rebate [31,32], while the waste-to-energy sector is eligible for a full rebate [33]. Furthermore, various short-term subsidy programs, often introduced prior to the standard feed-in tariffs, have been implemented to support renewable energy development. For example, from 2009 to 2013, the Golden Sun Demonstration Programme provided grants to large-scale photovoltaics projects, covering 50% of the cost of normal projects and 70% for projects in remote areas [34].

Official statistics show that these policies have resulted in a rapid growth in renewable energy, in particular wind and solar power. In 2005, the country had only 1260 MW of installed wind capacity, but by 2014, this capacity had increased to 95,810 MW, representing 31% of the total global wind capacity [35]. More recently, the solar photovoltaics capacity has been increasing at an equally rapid pace. In 2014, China's photovoltaics capacity had increased by 60% to 26,520 MW, with 4670 MW derived from distributed sources [36]. Overall, the share of renewable energy in the country's total installed capacity has increased from 23% in 2005 to 32% in 2014. However, the curtailment of renewable power, especially wind power, is becoming an endemic problem in China and a key barrier to renewable energy development. The National Energy Administration reported that 17,500 GWh of wind electricity was lost because of curtailment in the first half of 2015 alone [37]. Furthermore, up to 30% of the installed capacity was unable to obtain access to the grid [38]. In March 2015, the National Development and Reform Committee issued a series of instructions on methods to address the curtailment problem, including: prioritizing renewable energy in the annual electricity-balancing exercise, prioritizing imports of renewable energy for provinces where renewable energy is underdeveloped and providing financial support for regulating coal-fired power peaks [39]. The government intends to introduce renewable portfolio standards in the future, which will impose mandatory targets on local governments to ensure that a certain proportion of electricity consumption is derived from renewable energy sources [40].

The Chinese government has also devised three main strategies relevant to low-carbon energy, namely: (1) promoting nuclear energy; (2) supporting the use of natural gas; and (3) eliminating small and inefficient coal-fired units. Concerning nuclear energy, the release of the Mid- to Long-Term Development Plan for Nuclear Energy (2011–2020) in 2007 marked the beginning of a rapid development of nuclear power in the country. A number of tax incentives were introduced to support this, including a rebate on value-added tax (75% for the first five years of operation, 70% for Years 6–10 and 55% for Years 11–15) [41], as well as a rebate on import tax for essential equipment. Since nuclear power is cost competitive, the demand for this type of energy is high, especially in the coastal areas, where energy endowment is poor [42]. In light of the Fukushima accident in Japan, though, the central government of China has prioritized nuclear safety, and comprehensive inspections of nuclear facilities are therefore conducted [43].

With respect to natural gas, the intention of the government is to increase its use as a substitute for coal in the electricity sector and oil in the transport sector. Consequently, the government has recently started implementing a fast-track approval process for gas-fired units [44], while the Ministry of Transport has made funding available for liquefied natural gas projects for road and water transport [45].

China has numerous small and inefficient coal-fired power units, built during the 1980s and 1990s to address the severe electricity shortages at the time. These low-cost, but polluting units need to be replaced with renewable and low-carbon energy plants. In 1999, the central government issued a directive that coal-fired units smaller than 50 MW were to be decommissioned. However, this was not strictly implemented, and by 2006, the capacity of power units smaller than 10 MW was still more than 121 GW. A new policy was thus adopted in 2007 aiming to shut down: (1) coal-fired units smaller than 50 MW; (2) coal-fired units between 50 MW and 100 MW that had been in operation for 20 years or longer; and (3) coal-fired units between 100 MW and 200 MW that had exceeded their design life [46]. An innovative feature of this policy is bundling the decommissioning of dispensable plants with the approval for constructing new coal-fired power plants. For example, to receive approval from the central government to build a 1000-MW coal-fired power plant, the local authority is obliged to shut down at least 600 MW of small and inefficient units. The government also employs economic measures to achieve its goals, such as lowering the on-grid electricity tariffs of small coal-fired units [47]. This new policy has resulted in a wave of shutdowns, with over 76.8 GW of capacity being eliminated between 2006 and 2010 and a further 20 GW expected to be shut down between 2011 and 2015.

4. FLEXIBLE IMPLEMENTATION

Recent empirical studies have shed light on a highly complex and varied implementation process of clean energy programs in China. There is clear evidence that, contrary to the common perception of a highly hierarchical and top-down governance approach, the implementation process displays a high degree of flexibility [48,49]. This flexibility is revealed in policy targets, the amount of funding dedicated to implementation and the choice of policy instruments, scope and framing. Furthermore, it reflects the trend for decentralization of the policy process in the post-Mao period. While the central government remains the primary policymaker, it rarely involves itself in the implementation process, which is devolved to local governments. In fact, learning from the disastrous consequences of uniform implementation during the Mao period, the central government now encourages flexible implementation that is sensitive to local contexts.

This *de jure* flexibility coexists with what can be called *de facto* flexibility in China. Given the large size of the country, it is difficult for the central government to monitor local government behavior. While China's unitary (as opposed to federal) system gives the central government absolute power over local governments, the ability to exercise this power is constrained by the problem of asymmetric information and the cost involved in tackling it [50,51]. Conflicting priorities from the central government, for example between economic and environmental objectives, further weaken central control over local governments and enhance *de facto* flexibility in local implementation [52]. Therefore, even if the central government has the intention of tightening up the implementation process, it may lack the means and capability to do so.

While flexibility in the implementation process is a well-known Chinese phenomenon, there is an ongoing debate on whether, how much and what type of flexibility is conducive to desirable outcomes in terms of clean energy transition. Two contrasting viewpoints can be established. On the one hand, flexibility can improve policy implementation by allowing local governments to adapt to local contexts and stimulating policy innovation. There is some evidence to support this viewpoint: with regard to the retrofitting of existing buildings, the local government in Changchun changed the voluntary-based subsidy program to an essentially free, but involuntary retrofit program, resulting in the successful retrofitting of a large number of apartments in a very short period [53]. In another illustrative case, officials responsible for implementing the industrial energy-efficiency programs in Shanxi province were able to increase public support and minimize political resistance by policy-bundling: the programs were implemented as part of larger campaigns and connected with policies and concerns of a more local or immediate interest [54]. Furthermore, in Beijing, the government provides additional incentives for purchasing electric vehicles by exempting them from the strict car ownership control program.

On the other hand, flexible implementation can give local governments the opportunity to ignore, distort or undermine central directives, resulting in poor environmental outcomes compared to what the central government intended. There is equally compelling evidence to support this viewpoint, of which the implementation of the Ten Thousand Enterprises Programme in Changchun is an example. Due to cost-cutting in enforcement activities and the refusal to impose penalties on local enterprises that do not comply with the requirements, the local government has changed the mandatory energy-efficiency program into essentially a voluntary one, to the detriment of the program's performance [6]. Moreover, with regard to the elimination of backward-production capacity, many local governments have been caught falsifying the production capacity of the factories that were closed down or closing the same factory multiple times to claim extra credit.

Such conflicting views about China's flexible implementation of a clean energy policy cannot be adequately resolved without more empirical evidence and fine-grained analysis. In particular, several knowledge gaps must be addressed. First, there is a need to establish the nature and extent of flexibility in implementation: To what degree can local government deviate from central directives? In what ways? Where are the boundaries? How does the situation vary across different policy areas and programs? It is important to be mindful that implementation is a complex process, and there are many aspects of flexibility (targets, instruments, funding, *etc.*), which raises the question of which components should be made flexible. Past experience suggests that the central government typically allocates a high degree of flexibility in the amount of funding and policy instruments, but not with policy targets. Such an approach, however, has given rise to the phenomenon of meeting targets with methods that are detrimental to the long-term objectives. For example, in the latter half of 2010, many factories were shut down temporarily by local governments to reduce their emissions, without any long-term measures for energy-efficiency improvement.

Second, there is the form implementation strategies have taken in the context of flexible implementation: Do local governments adapt central directives to local contexts or, rather, use flexibility as an excuse for refusing to carry out their implementation responsibilities fully? What innovative strategies have been developed by local officials? Are they successful? Can the experience be transferred elsewhere? To answer these questions, we need empirical observation of the types of implementation strategies that have been adopted by local governments. It is also necessary to collect independent empirical evidence on the outcome of clean energy policy implementation. Many existing studies rely solely on official data, but such an over-reliance on government data is problematic, given the concerns about reliability and accuracy, including: evidence of data manipulation at both central and local levels, the lack of data in some crucial areas and the differences in terminologies and definitions [55]. Studies should not ignore the unintended consequences of implementing these policies either, in particular with regard to the people affected. For instance, shutting down small coal-fired units and backward-production capacity severely affects the livelihoods of a large number of people. Understanding the impact and how the government can assist with the transition are therefore important and valuable research endeavors.

Third, we need to explain the differences in approach to implementation, which involves asking such questions as: How do local officials determine implementation priorities? What are the factors and actors that shape such decisions? Why can a program be implemented well in one place, but not in another? Existing studies generally point to local interests as a crucial factor. The new role of local governments as a champion of local interests is a product of the post-Mao financial reforms that transferred greater financial responsibility to localities. Local officials now have great incentives to strengthen local economies to maintain a steady revenue. Moreover, intense competition between localities provides a huge motivation to local officials vying for promotion, which further strengthens the incentive to advance local interests. As a result, clean energy policies that are aligned more with local interests will receive more attention from the local government. While this argument is convincing, local interests are not the only determining factor, as there are many cases where local officials implement policies that ostensibly harm local interests. A more comprehensive explanation is needed, especially for the causes of poor implementation, which cannot be simply explained away by a lack of local interest. There is a need for a more systematic and multidimensional examination of the factors that shape the behavior of local officials, including local politics, institutional capacities and implementation challenges.

Fourth and finally, the way in which the central government governs the behavior of local government plays an important role in defining the flexibility of implementation. In particular, the role of the target responsibility system in clean energy transition must be better understood, as it has been shown to be an important instrument used by the central government to influence local implementation in the reform era [48]. In short, the target responsibility system establishes political incentives (e.g., promotion, annual honors) for local governments to implement certain programs or meet certain targets. How does the target responsibility system work in the context of clean energy transition? What kind of leverage can the central government apply to ensure compliance from local governments? What are the limitations? How can the central government overcome the asymmetric information problem that has limited its control over local governments? At what cost? More research is therefore needed to produce recommendations with regard to strengthening the ability of the central government to influence the implementation process.

5. CONCLUSIONS

In the context of China's rapid clean energy transition and its global implications, this paper has underscored a significant gap in knowledge generated by the emergence of numerous new clean energy policies and a dearth of empirical studies on their implementation. Existing studies have identified policy implementation and flexibility in the process as key features of and challenges to China's clean energy transition. However, questions regarding the nature and extent of flexibility in implementation, implementation strategies and their impacts, factors that shape the behavior of local officials responsible for implementation and the relationship between central-local relations and policy implementation need to be addressed. Answering these questions through empirical investigation can contribute to China's clean energy transition in many ways. At a local level, the findings can be used to generate context-specific recommendations for local governments to improve the implementation of clean energy policies. At the national level, the investigation can contribute to the debate on the relationship between flexible implementation and environmental outcomes, as well as help devise a governance system where the implementation process works toward, rather than against, China's important mission of clean energy transition.

ACKNOWLEDGMENTS

This research was supported by the Faculty Research Grant of Hong Kong Baptist University (FRG1/14-15/019).

CONFLICTS OF INTEREST

The authors declare no conflict of interest.

REFERENCES

1 State Council. Energy Development Strategic Plan (2014–2020). 2014. Available online: http://www.gov.cn/zhengce/content/2014-11/19/content_9222.htm (accessed on 25 August 2015).

2 National Development and Reform Commission. Enhanced Actions on Climate Change: China's Intended Nationally Determined Contributions. 2015. Available online: http://www4.unfccc. int/

submissions/INDC/Published%20Documents/China/1/China's%20INDC%20-%20on%2030%20June%202015.pdf (accessed on 25 August 2015).

3 Guan, D.; Liu, Z.; Geng, Y.; Lindner, S.; Hubacek, K. The gigatonne gap in China's carbon dioxide inventories. *Nat. Clim. Chang.* **2012**, *2*, 672–675.

4 Christoff, P. Cold climate in Copenhagen: China and the United States at COP15. *Environ. Polit.* **2010**, *19*, 637–656.

5 Conrad, B. China in Copenhagen: Reconciling the "Beijing climate revolution" and the "Copenhagen climate obstinacy". *China Q.* **2012**, *210*, 435–455.

6 Lo, K.; Li, H.; Wang, M. Energy conservation in China's energy-intensive enterprises: An empirical study of the Ten-Thousand Enterprises Program. *Energy Sustain. Dev.* **2015**, *27*, 105–111.

7 National Development and Reform Commission. Notice on Using Pricing Mechanisms to Promote the Structural Adjustment of the Cement Industry. 2014. Available online: http:// www.sdpc. gov.cn/fzgggz/jggl/zcfg/201405/t20140520_615511.html (accessed on 25 August 2015).

8 State Council. Opinion on Improving the Electricity Tariff Differentiation Policy. 2006. Available online: http:// www. gov.cn/zwgk/2006-09/22/content_396258.htm (accessed on 25 August 2015).

9 Beijing Municipal Commission of Development and Reform. Opinion on Improving the Electric Tariff Differentiation Policy in Beijing. 2015. Available online: http://www.bjpc.gov.cn/tztg/201507/t9232987. htm (accessed on 25 August 2015).

10 Ministry of Industry and Information Technology. Notice on the Elimination of Backward Production Capacity Implementation Plan. 2011. Available online: http://www.miit.gov.cn/n11293472/ n11293877/n13138101/n13138148/13583247.html (accessed on 25 August 2015).

11 National Development and Reform Committee. 11th FYP Energy Conservation Review: Significant Results for the Elimination of Backward Production Capacity Programme. 2011. Available online: http://www.gov.cn/gzdt/2011-03/10/content_ 1821724.htm (accessed on 25 August 2015).

12 Ministry of Housing and Urban-Rural Development. Notice on 2010 Building Energy Conservation Inspection Result. 2010. Available online: http://www.mohurd.gov.cn/zcfg/jsbwj_0/jsbwjjskj/201104/t20110421_203196.html (accessed on 25 August 2015).

13 Lee, W.; Chen, H. Benchmarking Hong Kong and China energy codes for residential buildings. *Energy Build.* **2008**, *40*, 1628–1636.

14 Zhou, N.; McNeil, M.; Levine, M. Assessment of building energy-saving policies and programs in China during the 11th five-year plan. *Energy Effic.* **2012**, *5*, 51–64.

15 Liu, Y.; Kokko, A. Who does what in China's new energy vehicle industry? *Energy Policy* **2013**, *57*, 21–29.

16 Ministry of Finance. Notice on 2016–2020 Fiscal Support Policy for New Energy Vehicles. 2015. Available online: http://jjs.mof.gov. cn/ zhengwuxinxi/zhengcefagui/201504/t20150429_1224515.html (accesse d on 25 August 2015).

17 Ministry of Finance. Notice on the Vehicle and Vessel Tax Concession for Energy Efficient and New Energy Vehicles and Vessels. 2015. Available online: http://www.chinatax.gov.cn/ n810341/n810755/ c1617349/content.html (accessed on 25 August 2015).

18 Xinhua News Agency. China's New Energy Vehicles Ownership Passed 120,000. 2015. Available online: http://news.xinhuanet.com/ finance/ 2015-03/12/c_1114622428.htm (accessed on 25 August 2015). [Google Scholar]

19 Lo, K. Interested but unsure: Public attitudes toward electric vehicles in China. *Electron. Green J.* **2013**, *36*, 1–12.

20 Standardization Administration. GB 19578-2014 Fuel Consumption Limits for Passenger Cars. 2014. Available online:http://gb123. sac.gov.cn/gb/showGb?gbCode=GB%2019578-2014 (accessed on 25 August 2015).

21 National Development and Reform Commission. Mid- to Long-Term Development Plan for Renewable Energy. 2007. Available online: http://www.china.com.cn/policy/txt/2007-09/04/content_ 8800358. htm (accessed on 25 August 2015).

22 Lema, A.; Ruby, K. Between fragmented authoritarianism and policy coordination: Creating a Chinese market for wind energy. *Energy Policy* **2007**, *35*, 3879–3890.

23 National Development and Reform Commission. Notice on Optimising Wind Power Feed-in Tariff Policy. 2009. Available online: http:// www.sdpc.gov.cn/zwfwzx/zfdj/jggg/dian/200907/t20090727_292837.ht ml (accessed on 25 August 2015).

24 BeiJiXing. China's Key Wind Power Developers. 2014. Available online: http://news.bjx.com.cn/html/20141212/572965.shtml (accessed on 25 August 2015).

25 National Development and Reform Commission. Notice on Improving Agriculture and Forestry Feed-in Tariff Policy. 2010. Available online: http://www.ndrc.gov.cn/zwfwzx/zfdj/jggg/dian/201007/t20100728_363366.html (accessed on 25 August 2015).

26 National Development and Reform Commission. Notice on Offshore Wind Power Feed-in Tariff Policy. 2014. Available online: http://www.sdpc.gov.cn/zwfwzx/zfdj/jggg/dian/201406/t20140619_615708.html (accessed on 25 August 2015).

27 Shanghai Municipal Development and Reform Commission. Shanghai Renewable and New Energy Development Support Fund. 2014. Available online: http://www.shdrc.gov.cn/ main?main_colid= 319&top_id=312&main_artid=24296 (accessed on 25 August 2015).

28 Ministry of Finance. Notice on Adjusting Renewable Energy Surcharge Standard. 2013. Available online: http://zhs.mof.gov.cn/ zhengwuxinxi/zhengcefabu/201310/t20131009_996652.html (accessed on 25 August 2015).

29 Ministry of Finance; State Administration of Taxation. Notice on Utility-Scale Hydroelectricity Value-Added Tax Policy. 2014. Available online: http://202.108.90.131/SuniT/202.108.90.130/n810341/n810755/c1149889/content.html(accessed on 25 August 2015).

30 Ministry of Finance; State Administration of Taxation. Notice on Simplifying Value-Added Tax Policy. 2014. Available online: http://www.chinatax.gov.cn/n810341/n810755/c1150435/content.html (accessed on 25 August 2015).

31 Ministry of Finance; State Administration of Taxation. Notice on Photovoltaics Power Value-Added Tax Policy. 2013. Available online: http://www.chinatax.gov.cn/n810341/n810765/n812146/n812323/c1080750/content.html (accessed on 25 August 2015).

32 Ministry of Finance; State Administration of Taxation. Notice on Wind Power Value-Added Tax Policy. 2015. Available online: http://szs.mof.gov.cn/zhengwuxinxi/zhengcefabu/201506/t20150616_1256851.html (accessed on 25 August 2015).

33 Ministry of Finance; State Administration of Taxation. Notice on Resources Utilisation and Other Products Value-Added Tax Policy. 2008. Available online: http://www.chinatax.gov.cn/ n810341/n810765/

n812171/n812675/c1190512/content.html (accessed on 25 August 2015).

34 Ministry of Finance; Ministry of Science and Technology; National Energy Administration. Notice on the Implementation of the Golden Sun Demonstration Programme. 2009. Available online: http:// big5.gov.cn/gate/big5/www.gov.cn/zwgk/2009-07/21/content_ 1370811.htm (accessed on 25 August 2015).

35 *Global Wind Report Annual Market Update 2014*; Global Wind Energy Council: Brussels, Belgium, 2015; Available online: http:// www. gwec.net/wp-content/uploads/ 2015/03/GWEC_ Global_ Wind_ 2014_ Report_LR.pdf (accessed on 25 August 2015).

36 National Energy Administration. Statistics of Photovoltaics Power in 2014. 2015. Available online: http://www.nea.gov.cn/2015-03/09/c_ 134049519.htm (accessed on 25 August 2015).

37 National Energy Administration. Operation of Wind Power in the First Half of 2015. 2015. Available online: http://www.nea.gov.cn/2015-07/27/c_134451678.htm (accessed on 25 August 2015).

38 Zhao, X.; Zhang, S.; Yang, R.; Wang, M. Constraints on the effective utilization of wind power in China: An illustration from the northeast China grid. *Renew. Sustain. Energy Rev.* **2012**, *16*, 4508–4514.

39 National Development and Reform Commission. Opinions on Improving the Balancing of Electricity to Achieve Full Utilisation of Renewable Energy. 2015. Available online: http://www.sdpc.gov. cn/gzdt/201503/t20150323_668207.html(accessed on 25 August 2015).

40 China Energy News. Renewable Energy Portfolio Almost Ready. 2015. Available online: http://paper.people.com.cn/zgnyb/html/2015-04/20/ content_1556462.htm (accessed on 25 August 2015).

41 Ministry of Finance; State Administration of Taxation. Notice on Taxation Policy of the Nuclear Power Industry. 2008. Available online: http://www.chinatax.gov.cn/n810341/n810755/c1224064/conten t.html (accessed on 25 August 2015).

42 Xu, Y. The struggle for safe nuclear expansion in China. *Energy Policy* **2014**, *73*, 21–29.

43 Mu, R.; Zuo, J.; Yuan, X. China's approach to nuclear safety—From the perspective of policy and institutional system.*Energy Policy* **2015**, *76*, 161–172.

44 National Development and Reform Commission; National Energy Administration; Ministry of Housing and Urban-Rural Development.

Notice on Natural-Gas Distributed Power Demonstration Project Implementation Detail. 2014. Available online: http:// www.chinade.org.cn/upload/accessory/201411/20141151536383771275.pdf (accessed on 25 August 2015).

45 Ministry of Transport. Instruction on Advancing the Use of Liquefied Natural Gas in Water Transportation. 2013. Available online: http://www.mot.gov.cn/zfxxgk/bnssj/syj/201311/t20131104_1507474.html (accessed on 25 August 2015).

46 State Council. Opinion on Expediating the Elimination of Small Coal-Fired Units. 2007. Available online: http://www.sdpc.gov. cn/zcfb/zcfbqt/200701/t20070131_115037.html (accessed on 25 August 2015).

47 National Development and Reform Commission. Notice on Lowering On-Grid Electricity Tariff of Small Coal-Fired Power Units. 2007. Available online: http://www.sdpc.gov.cn/ zwfwzx/zfdj/jggg/dian/200704/t20070405_128800.html(accessed on 25 August 2015).

48 Lo, K. China's low-carbon city initiatives: The implementation gap and the limits of the target responsibility system.*Habitat Int.* **2014**, *42*, 236–244.

49 He, J. Governing forest restoration: Local case studies of sloping land conversion program in southwest China. *For. Policy Econ.* **2014**, *46*, 30–38.

50 Zheng, Y. *De Facto Federalism in China: Reforms and Dynamics of Central-Local Relations*; World Scientific Publishing Company: Hackensack, NJ, USA, 2007.

51 Li, L.C. Central-local relations in the People's Republic of China: Trends, processes and impacts for policy implementation. *Public Adm. Dev.* **2010**, *30*, 177–190.

52 Ran, R. Perverse incentive structure and policy implementation gap in China's local environmental politics. *J. Environ. Policy Plan.* **2013**, *15*, 17–39.

53 Lo, K. The "warm houses" program: Insulating existing buildings through compulsory retrofits. *Sustain. Energy Technol. Assess.* **2015**, *9*, 63–67.

54 Kostka, G.; Hobbs, W. Local energy efficiency policy implementation in China: Bridging the gap between national priorities and local interests. *China Q.* **2012**, *211*, 765–785.

55 Sinton, J.E. Accuracy and reliability of China's energy statistics. *China Econ. Rev.* **2001**, *12*, 372–383.

CHAPTER 4

Energy and sustainable development in Nigeria: The way forward

Sunday Olayinka Oyedepo

Mechanical Engineering Department, Covenant University

ABSTRACT

Access to clean modern energy services is an enormous challenge facing the African continent because energy is fundamental for socioeconomic development and poverty eradication. Today, 60% to 70% of the Nigerian population does not have access to electricity. There is no doubt that the present power crisis afflicting Nigeria will persist unless the government diversifies the energy sources in domestic, commercial, and industrial sectors and adopts new available technologies to reduce energy wastages and to save cost. This review examines a set of energy policy interventions, which can make a major contribution to the sustainable economic, environmental, and social development of Africa's most populated country, Nigeria. Energy efficiency leads to important social benefits, such as reducing the energy bills for poor households. From an economic point of view, implementing the country's renewable energy target will have significant costs, but these can partly be offset by selling carbon credits according to the rules of the 'Clean Development Mechanism' agreed some 10 years ago, which will result in indirect health benefits.

Nigeria could benefit from the targeted interventions that would reduce the local air pollution and help the country to tackle greenhouse gas emissions. Many factors that need to be considered and appropriately addressed in the shift to its sustainable energy future are examined in this article. These include a full exploitation and promotion of renewable energy resources, energy efficiency practices, as well as the application of energy conservation measures in various sectors such as in the construction of industrial, residential, and office buildings, in transportation, etc.

KEYWORDS

Sustainable energy, Renewable energy, Energy efficiency, Energy conservation

1. BACKGROUND

Energy plays the most vital role in the economic growth, progress, and development, as well as poverty eradication and security of any nation. Uninterrupted energy supply is a vital issue for all countries today. Future economic growth crucially depends on the long-term availability of energy from sources that are affordable, accessible, and environmentally friendly. Security, climate change, and public health are closely interrelated with energy[1]. Energy is an important factor in all the sectors of any country's economy. The standard of living of a given country can be directly related to the *per capita* energy consumption. The recent world's energy crisis is due to two reasons: the rapid population growth and the increase in the living standard of whole societies. The *per capita* energy consumption is a measure of the *per capita*income as well as a measure of the prosperity of a nation[2].

Energy supports the provision of basic needs such as cooked food, a comfortable living temperature, lighting, the use of appliances, piped water or sewerage, essential health care (refrigerated vaccines, emergency, and intensive care), educational aids, communication (radio, television, electronic mail, the World Wide Web), and transport. Energy also fuels productive activities including agriculture, commerce, manufacturing, industry, and mining. Conversely, a lack of access to energy contributes to poverty and deprivation and can contribute to the economic decline. Energy and poverty reduction are not only closely connected with each other, but also with the socioeconomic development, which involves productivity, income growth, education, and health[3].

The energy crisis, which has engulfed Nigeria for almost two decades, has been enormous and has largely contributed to the incidence of poverty by paralyzing industrial and commercial activities during this period. The Council for Renewable Energy of Nigeria estimates that power outages brought about a loss of 126 billion naira (US$ 984.38 million) annually[4].

Apart from the huge income loss, it has also resulted in health hazards due to the exposure to carbon emissions caused by constant use of 'backyard generators' in different households and business enterprises, unemployment, and high cost of living leading to a deterioration of living conditions.

Moreover, according to the Central Bank estimate in 1985, Nigeria consumed 8,771,863 tonnes of oil equivalent[5]. This is equal to about 180,000 barrels of oil per day. Since then, oil consumption in Nigeria has drastically increased. The effect of this increase on the economy relying solely on revenue from oil is tremendous. Also, the Department for Petroleum Resources[6] reported an amount of petroleum of more than 78% of the total energy consumption in Nigeria. In the present predicament as a nation, it is obvious that depending mainly on fossil fuel (petroleum) is not enough to meet the energy needs of the country. Since Nigeria is blessed with abundant renewable energy resources such as hydroelectric, solar, wind, tidal, and biomass, there is a need to harness these resources and chart a new energy future for Nigeria. In this regard, the government has a responsibility to make renewable energy available and affordable to all.

Many indigenous researchers have looked into the availability of renewable energy resources in Nigeria with a view to establishing their viability in the country. Onyebuchi[7] estimated the technical potential of solar energy in Nigeria with a 5% device conversion efficiency put at 15.0×10^{14} kJ of useful energy annually. This equates to about 258.62 million barrels of oil equivalent annually, which corresponds to the current national annual fossil fuel production in the country. This will also amount to about 4.2×10^5 GW/h of electricity production annually, which is about 26 times the recent annual electricity production of 16,000 GW/h in the country. In their work, Chineke and Igwiro[8] show that Nigeria receives abundant solar energy that can be usefully harnessed with an annual average daily solar radiation of about 5.25 kW h/m^2/day. This varies between 3.5 kW h/m^2/day at the coastal areas and 7 kW h/m^2/day at the northern boundary. The average amount of sunshine hours all over the country is estimated to be about 6.5 h. This gives an average annual solar energy intensity of 1,934.5 kW h/m^2/year; thus, over the course of a year, an average of 6,372,613 PJ/year (approximately 1,770 TW h/year) of solar energy falls on the entire land area of Nigeria. This is about 120,000times the total annual average electrical energy generated by the Power Holding Company of Nigeria (PHCN). With a 10% conservative conversion efficiency, the available solar energy resource is about 23 times the Energy Commission of Nigeria's (ECN) projection of the total final energy demand for Nigeria in the year 2030[9]. To enhance the developmental trend in the country, there is every need to support the existing unreliable energy sector with a sustainable source of power supply through solar energy.

Moreover, many indigenous researchers have also explored the availability of wind energy sources in Nigeria with a view of implementing them if there is a likelihood for their usage. Adekoya and Adewale[10] analyzed the wind speed data of 30 stations in Nigeria, determining the annual mean wind speeds and power flux densities, which vary from 1.5 to 4.1 m/s to 5.7 to 22.5 W/m^2, respectively. Fagbenle and Karayiannis[11] carried out a 10-year wind data analysis from 1979 to 1988, considering the surface and upper winds as well as the maximum gusts, whereas Ngala et al.[12] performed a statistical analysis of the wind energy potential in Maiduguri, Borno State, using the Weibull distribution and 10-year (1995 to 2004) wind data. A cost benefit analysis was also performed using the wind energy conversion systems for electric power generation and supply in the State. Each of these reports point to the fact that the nation is blessed with a vast opportunity for harvesting wind for electricity production, particularly at the core northern states, the mountainous parts of the central and eastern states, and also the offshore areas, where wind is abundantly available throughout the year. The issue then is for the country to look at ways of harnessing resources towards establishing wind farms in various regions and zones that have been identified as possessing abilities for the harvesting of wind energy.

Akinbami[13] reported that the total hydroelectric power potential of the country was estimated to be about 8,824 MW with an annual electricity generation potential in excess of 36,000 GW h. This consists of 8,000 MW of large hydropower technology, while the remaining 824 MW is still small-scale hydropower technology. Presently, 24% and 4% of both large and small hydropower potentials, respectively, in the country have been exploited.

Akinbami et al.'s assessment[14] indicated that the identified feedstock substrate for an economically feasible biogas program in Nigeria includes water lettuce, water hyacinth, dung, cassava leaves, urban refuse, solid (including industrial) waste, agricultural residues, and sewage. The authors' views include the following: Nigeria produces about 227,500 tonnes of fresh animal wastes daily. Since 1 kg of fresh animal wastes produces about 0.03 m^3 gas, then Nigeria could produce about 6.8 million m^3 of biogas every day. In addition to all these, 20 kg of municipal solid wastes per capital has been estimated to be generated in the country annually.

The prime objectives of this paper are (1) to review the current status of the energy resources, the energy demand, and supply in Nigeria and (2) to explore the prospects of utilizing renewable energy resources and to increase the energy efficiency as a possible means of sustainable development in Nigeria.

1.1 Energy situation in Nigeria

Nigeria is Africa's energy giant. It is the continent's most prolific oil-producing country, which, along with Libya, accounts for two-thirds of Africa's crude oil reserves. It ranks second to Algeria in natural gas[15]. Most of Africa's bitumen and lignite reserves are found in Nigeria. In its mix of conventional energy reserves, Nigeria is simply unmatched by any other country on the African continent. It is not surprising therefore that energy export is the mainstay of the Nigerian economy. Also, primary energy resources dominate the nation's industrial raw material endowment.

Several energy resources are available in Nigeria in abundant proportions. The country possesses the world's sixth largest reserve of crude oil. Nigeria has an estimated oil reserve of 36.2 billion barrels. It is increasingly an important gas province with proven reserves of nearly 5,000 billion m^3. The oil and gas reserves are mainly found and located along the Niger Delta, Gulf of Guinea, and Bight of Bonny. Most of the exploration activities are focused in deep and ultra-deep offshore areas with planned activities in the Chad basin, in the northeast. Coal and lignite reserves are estimated to be 2.7 billion tons, while tar sand reserves represent 31 billion barrels of oil equivalent. The identified hydroelectricity sites have an estimated capacity of about 14,250 MW. Nigeria has significant biomass resources to meet both traditional and modern energy uses, including electricity generation[16]. Table1 shows Nigeria's energy reserves/potentials. There has been a supply and demand gap as a result of the inadequate development and inefficient management of the energy sector. The supply of electricity, the country's most used energy resource, has been erratic[17].

Table 1 Nigeria's energy reserves/capacity as in December 2005

Resource type	**Reserves**	**Reserves (BTOE)[c]**	**Reserves (× 10^7) TJ**
Crude oil	36.2 billion barrels	4.896	20.499
Natural gas	166 trillion SCF[a]	4.465	18.694
Coal and lignite	2.7 billion tonnes	1.882	7.879
Tar sands	31 billion barrels of oil equivalent	4.216	17.652

Resource type	Reserves	Reserves (BTOE)[c]	Reserves (× 10^7) TJ
Subtotal Fossil		15.459	64.724
Hydropower, large Scale	11,000 MW		0.0341/year
Hydropower, small Scale	3,250 MW		0.0101/year
Fuel wood	13,071,464 ha[b]		
Animal waste	61 million tonnes/year		
Crop residue	83 million tonnes/year		
Solar radiation	3.5 to 7.0 kW h/m^2/day		
Wind	2 to4 m/s (annual average) at 10 m in height		

[a]SCF, standard cubic feet; [b]forest land estimate for 1981; [c]BTOE, billion tonnes of oil equivalent. Adapted from ECN[18].

The situation in the rural areas of the country is that most end users depend on fuel wood. Fuel wood is used by over 70% of Nigerians living in the rural areas. Nigeria consumes over 50 million tonnes of fuel wood annually, a rate which exceeds the replenishment rate through various afforestation programs. Sourcing fuel wood for domestic and commercial uses is a major cause of desertification in the arid-zone states and erosion in the southern part of the country. The rate of deforestation is about 350,000 ha/year, which is equivalent to 3.6% of the present area of forests and woodlands, whereas reforestation is only at about 10% of the deforestation rate[19].

The rural areas, which are generally inaccessible due to the absence of good road networks, have little access to conventional energy such as electricity and petroleum products. Petroleum products such as kerosene and gasoline are purchased in the rural areas at prices 150% in excess of their official pump prices. The daily needs of the rural populace for heat energy are therefore met almost entirely from fuel wood. The sale of fuel wood and charcoal is mostly uncontrolled in the unorganized private sector. The sale of kerosene, electricity and cooking gas is essentially influenced and controlled by the Federal Government or its agencies - the Nigerian National Petroleum Corporation (NNPC) in the case of kerosene and cooking gas, and the PHCN in the case of electricity. The policy of the Federal Government had been to subsidize the pricing of locally consumed petroleum products, including electricity. In a bid to make the petroleum downstream sector more efficient and in an attempt to stem petroleum product consumption as a policy focus, the government has reduced and removed subsidies on various energy resources in Nigeria. The various policy options have always engendered price increases of the products[20].

With the restructuring of the power sector and the imminent privatization of the electricity industry, it is obvious that for logistic and economic reasons especially in the privatized power sector, rural areas that are remote from the grid and/or have low consumption or low power purchase potential will not be attractive to private power investors. Such areas may remain unserved into the distant future[21].

Meanwhile, electricity is required for such basic developmental services as pipe borne water, health care, telecommunications, and quality education. The poverty eradication and Universal Basic Education programs require energy for success. The absence of reliable energy supply has not only left the rural populace socially backward, but has also left their economic potentials untapped. Fortunately, Nigeria is blessed with abundant renewable energy resources such as solar, wind, biomass, and small hydropower potentials. The logical solution is increased penetration of renewables into the energy supply mix[15].

1.2 Energy consumption pattern in Nigeria

Energy consumption patterns in the world today shows that Nigeria and indeed African countries have the lowest rates of consumption. Nevertheless, Nigeria suffers from an inadequate supply of usable energy due to the rapidly increasing demand, which is typical of a developing economy. Paradoxically, the country is potentially endowed with sustainable energy resources. Nigeria is rich in conventional energy resources, which include oil, national gas, lignite, and coal. It is also well endowed with renewable energy sources such as wood, solar, hydropower, and wind[17].

The patterns of energy usage in Nigeria's economy can be divided into industrial, transport, commercial, agricultural, and household sectors[22]. The household sector accounts for the largest share of energy usage in the country - about 65%. This is largely due to the low level of development in all the other sectors.

The major energy-consuming activities in Nigeria's households are cooking, lighting, and use of electrical appliances. Cooking accounts for a staggering 91% of household energy consumption, lighting uses up to 6%, and the remaining 3% can be attributed to the use of basic electrical appliances such as televisions and pressing irons[9].

The predominant energy resources for domestic and commercial uses in Nigeria are fuel wood, charcoal, kerosene, cooking gas and electricity[20]. Other sources, though less common, are sawdust, agricultural crop residues of corn stalk, cassava sticks, and, in extreme cases, cow dung. In Nigeria, among the urban dwellers, kerosene and gas are the major cooking fuels. The majority of the people rely on kerosene stoves for domestic cooking, while only a few use gas and electric cookers[23].

The rural areas have little access to conventional energy such as electricity and petroleum products due to the absence of good road networks. Petroleum products such as kerosene and gasoline are purchased in the rural areas at prices very high in excess of their official pump prices. The rural population, whose needs are often basic, therefore depends to a large extent on fuel wood as a major traditional source of fuel. It has been estimated that about 86% of rural households in Nigeria depend on fuel wood as their source of energy[24]. A fuel wood supply/demand imbalance in some parts of the country is now a real threat to the energy security of the rural communities[22].

The energy consumption *per capita* in Nigeria is very small - about one-sixth of the energy consumed in developed countries. This is directly linked to the level of poverty in the country. Gross domestic product (GDP) and *per capita* income are indices that are used to measure the economic well-being of a country and its people[25]. GDP is defined as the total market value of all final goods and services produced within a given country in a given period of time (usually a calendar year). The *per capita* income refers to how much each individual receives, in monetary terms, of the yearly income that is generated in his/her country through productive activities. That is what each citizen would receive if the yearly income generated by a country from its productive activities were divided equally between everyone.

1.3 Current electricity situation in Nigeria

The electricity system in Nigeria centers on PHCN, which accounts for about 98% of the total electricity generation[26]. Power generation by other agencies such as the Nigerian Electricity Supply Company relies on thermal power for electricity generation unlike PHCN, which relies on both hydro- and thermal power. However, electricity is also a consumer of fuel and energy such as fuel oil, natural gas, and diesel oil. The importance of these sources of energy and fuel for generating electricity has been decreasing in recent years. However, hydropower that is relatively cheaper than these sources has grown to be more important than other sources[27]. However, more recently, the Power Authority has generated electricity through a mix of both thermal and hydro systems. All the power, distribution, and substations are specially interlinked by a transmission network popularly known as the national grid. The entire electricity generated nationwide is pooled into the National Control Centre, Osogbo, from where electricity is distributed to all parts of Nigeria.

The national electricity grid presently consists of 14 generating stations (3 hydro and 11 thermal) with a total installed capacity of about 8,039 MW as shown in Table2. The transmission network is made up of 5,000 km of 330-kV lines, 6,000 km of 132-kV lines, 23 of 330/132-kV substations, with a combined capacity of 6,000 or 4,600 MVA at a utilization factor of 80%. In turn, the 91 of 132/33-kV substations have a combined capacity of 7,800 or 5,800 MVA at a utilization factor of 75%. The distribution sector is comprised of 23,753 km of 33-kV lines, 19,226 km of 11-kV lines, and 679 of 33/11-kV substations. There are also 1,790 distribution transformers and 680 injection substations[28]. Table2 shows a summary of the generation capabilities of PHCN power stations as operated in the year 2008 (January to December)[29].

Table 2 Summary of generation capabilities of PHCN power stations as operated in 2008 (January to December)

Plant	Operator	Age (year)	Type	Installed capacity (MW)	Average availability (MW)	Availability factor	Number of units installed	Current number available
Kainji	PHCN	38 to 40	Hydro	760	438.86	0.58	8	6
Jebba	PHCN	25	Hydro	578.4	529.40	0.92	6	4
Shiroro	PHCN	22	Hydro	600	488.82	0.81	4	4
Egbin	PHCN	23	ST	1320	694.97	0.53	6	5
AES	AES	7	GT	315	233.91	0.77	9	9
Ajaokuta	STS	NA	GT	110	24.88	0.23	2	2
Sapele	PHCN	26 to 30	ST/GT	1020	156.60	0.15	10	1
Okpai	AGIP	3	GT/ST	480	394.56	0.88	3	3
Afam	PHCN	8 to 45	GT	709.6	82.12	0.09	20	3
Delta	PHCN	18	GT	912	211.67	0.24	18	12
Geregu	PHCN	NA	GT	414	305.14	0.74	3	3
Omoku	RS	3	GT	150	87.27	0.87	6	4
Omotosho	PHCN	1	GT	335	256.58	0.77	8	2
Olorunsogo	PHCN	1	GT	335	271.46	0.81	8	2
Total				8,039	4176.24	0.50	93	45

Adapted from PHCN [30].

As it can be seen in Table2, the existing plants operate at far below their installed capacity as many of them have units that need to be rehabilitated, retrofitted, and upgraded[31]. The percentage of generation capability from hydro turbines is 34.89%; from gas turbine, 35.27%; and from steam turbines, 29.84%. The relative contribution of the hydropower stations to the total electricity generation (megawatt per hour) is greater than that of the thermal power stations.

In terms of the consumption of electricity, a classification into three groups has been proposed (industrial, residential, and street light consumption). In 1970, the total electricity consumption stood at 145.3 MW/h; this increased to about 536.9 MW/h in 1980. However, in 2005, the total electricity consumption had increased to 1,873.1 MW/h[32]. On the generation side, these values of 176.6 MW/h in 1970 increased to 815.1 MW/h in 1980. By the end of 2005, the achieved total electricity generation was 2,997.3 MW/h[32]. Comparing the *per capita* power generation to that of other countries, Nigeria has the lowest among the countries, as shown in Table3, while the USA has the highest *per capita* electricity generation.

Table 3 Country statistics of electricity generation and ***per capita*** consumption

Continent	Country	Population (million)	Generation capacity (MW)	*Per capita* consumption (kW)
North America	USA	250	813,000	3.2
South America	Cuba	10.54	4,000	0.38
Europe (central)	UK	57.5	76,000	1.1
Europe (eastern)	Ukraine	49	54,000	1.33
Middle East	Iraq	23.6	10,000	0.42
Far East	South Korea	47	52,000	1.10
Africa	Nigeria	140	<4,000	0.03
	Egypt	67.9	18,000	0.27
	South Africa	44.3	45,000	1.02

Adapted from Okafor and Joe-Uzuegbu [17].

In spite of the contribution of electricity to the total gross domestic product, it is evident that Nigeria is facing several problems. The incapacity of the electricity subsector to efficiently meet the demand for electricity in the country has been caused by a number of problems, which have been detrimental to economic growth. The Central Bank of Nigeria[26] has identified nine problems associated with the National Electric Power Authority (NEPA) (now PHCN):

- Lack of preventive and routine maintenance of NEPA's facilities, resulting in huge energy losses.
- Frequent major breakdowns, arising from the use of outdated and heavily overloaded equipment.
- Lack of coordination between town planning authorities and PHCN, resulting in poor overall power system planning and overloading of PHCN equipment.
- Inadequate generation due to operational/technical problems arising from machine breakdown, low gas pressure, and low water levels.
- Poor funding of the organization.
- Inadequate budgetary provision and undue delay in release of funds to PHCN.
- PHCN's inefficient billing and collection system.
- High indebtedness to PHCN by both public and private consumers who are reluctant to pay for electricity consumed when due.
- Vandalizing and pilfering of PHCN equipment.

In addition to these, most of the existing electricity plants in Nigeria are underutilized or not functioning at all. Numerous reasons could be sighted as responsible for the underutilization of these plants. Some of which are (1) scarcity of relevant manpower for adequate maintenance and general consumer indiscipline; (2) lack of essential spare parts for maintenance of the plants; (3) absence of local manufacturing capabilities;(4) lack of systematic studies of distribution networks to reduce the extraordinary losses that usually accompany haphazard system expansion; and (5) inability to convert gas flares to a source of electricity[33].

The inefficiency as well as the inadequate facilities to boost electricity supply also have been major causes of the increasing gap between the demand and the supply of electricity. This could be due to the fact that there are only 14 generating stations in Nigeria (3 hydro and 11 thermal stations). Out of the approximated 8,039 MW of installed capacity in Nigeria, not more than 4,500 MW is ever produced. This is due to poor maintenance, fluctuation in water levels powering the hydro plants, and the loss of electricity in transmission. It could also be due to the 80-MW export of electricity each to the republic of Niger and Benin. ‘Apart from serving as a pillar of wealth creation in Nigeria, electricity is also the nucleus of operations and subsequently the engine of growth for all sectors of the economy'[34]. It has been indirectly re-echoed that electricity consumption is positively related to economic growth and that the former is a causal factor of the latter. This means that electricity consumption has diverse impacts on a range of socioeconomic activities and consequentially the living standards of Nigerians.

Notwithstanding the above pitfalls that had rendered public electricity supply in Nigeria unreliable and inefficient, the trend of its utilization has grown significantly over the past years. Figure1 shows the total electricity consumption in megawatts per hour and the various sectorial decompositions. Electricity utilization by the industrial sector has been fairly static because of the unreliable nature of the public electricity supply system in the country. Thus, many companies have resolved to provide their own power-generating sets as sources of electricity, leading to huge transfer costs on their products and services.

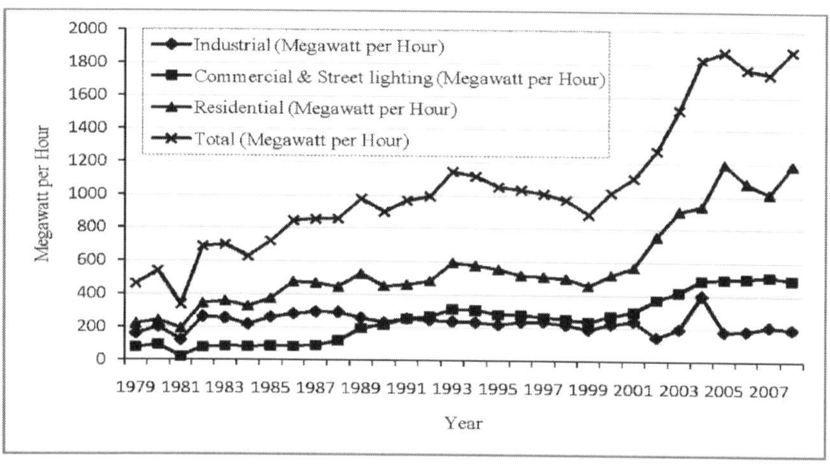

Figure 1 Electricity consumption pattern in Nigeria. **Adapted from CBN[35].**

Studies and experiences have shown that power generation in the country has been dismal and unable to compare with what has been obtained in smaller African countries. Manufacturers Association of Nigeria (MAN) gave the following performance indicators in Table4 for Nigeria's electricity sector compared with those of some other countries[28]. The data for some Southern Africa Development Community (SADC) countries such as Botswana and South Africa are comparable to those of the USA and France. The performance of the Nigerian power sector on the International Best Practices comparative rating is disgraceful. Perhaps, no other sector feels it as much as the manufacturing industrial sector wherein some notable international companies and organizations are on self-generated electricity 24 h/day for the 365 days of each year, as confirmed by the United Nations Industrial Development Organization in 2009[36]. The survey showed that, on average, manufacturers generated about 72% of the total power required to run their factories.

Table 4 Power supply reliability indices (international best practices)

Index	USA	Singapore	France	Nigeria (NEPA data)	Nigeria (MAN study)
SAIDI[a](min)	88	1.5	52	900	≥60,000
SAIFI[b](number/year	1.5	NA	NA	5	≥600
CAIDI[c](h)	0	NA	0	9	15
ASAI[d]	1	1	1	NA	≤0.4

[a]SAIDI, System average international duration index – Annual average total duration of power interruption to a consumer, in minutes; [b]SAIFI, System average interruption frequency index – Average number of interruptions of supply that a consumer experiences annually; [c]CAIDI, Consumer average interruption duration index – Average duration of an interruption of supply for a consumer who experiences the interruption on an annual basis, in hours; [d]ASAI, Average service availability index – Ratio of (Consumer hours service availability)/(Consumer hours service demanded). Adapted from Fagbenle et al.[28].

1.4 The Nigerian energy challenge

Nigeria's energy need is on the increase, and its increasing population is not adequately considered in the energy development program. The present urban-centered energy policy is deplorable, as cases of rural and sub-rural energy demand and supply do not reach the center stage of the country's energy development policy. People in rural areas depend on burning wood and traditional biomass for their energy needs, causing great deforestation, emitting greenhouse gases, and polluting the environment, thus creating global warming and environmental concerns. The main task has been to supply energy to the cities and various places of industrialization, thereby creating an energy imbalance within the country's socioeconomic and political landscapes. Comparing the present and ever increasing population with the total capacity of the available power stations reveals that Nigeria is not able to meet the energy needs of the people. The rural dwellers still lack electric power[37].

The nature of Nigeria's energy crises can be characterized by two key factors. The first concerns the recurrent severe shortages of the petroleum product market of which kerosene and diesel are the most prominent. Nigeria has five domestic refineries owned by the government with a capacity to process 450,000 barrels of oil per day, yet imports constitute more than 75% of petroleum product requirements. The state-owned refineries have hardly operated above a 40% capacity utilization rate for any extended period of time in the past two decades. The gasoline market is much better supplied than kerosene and diesel because of its higher political profile. This factor explains why the government has embarked on large import volumes to remedy domestic shortages of the product. According to the Minister for Energy, the subsidy to support the imports of gasoline alone will be in the range of 700 to 800 billion naira in 2008[38]. The weaker political pressures exerted by the consumers of kerosene (the poor and low middle class) and diesel (industrial sector) on the government and the constraints on public financing of large-scale imports of these products, as in the case of gasoline, largely explain their more severe and persistent market shortages[39].

The second dimension of Nigeria's energy crises is exemplified by such indicators as electricity blackouts, brownouts, and pervasive reliance on self-generated electricity. This development has occurred despite abundant energy resources in Nigeria. The electricity market, dominated on the supply side by the state-owned PHCN, formerly called NEPA, has been incapable of providing minimum acceptable international standards of electricity service reliability, accessibility, and availability for the past three decades[40]. The nature of this poor record in electricity supply is apparent in the trend in transmission and distribution losses shown in Figure2. The double-digit transmission and distribution losses are extremely large by international standards and are among the highest in the world. The system losses are five to six times higher than those in well-run power systems. The high level of power losses and the significant illegal access to the public power supply are indicative of the crisis in the industry.

Though the peak electricity demand has been less than half of the installed capacity in the past decade, load shedding occurs regularly. Power outages in the manufacturing sector provide another dimension to the crisis. In 2004, the major manufacturing firms experienced 316 outages. This increased by 26% in 2005, followed by an explosive 43% increase between 2006 and 2007. Though no published data exist, the near collapse of the generating system to far below 2,000 MW for prolonged periods of time suggests a reason for the number of outages in 2008 to be very high. This poor service delivery has rendered public supply a standby source as many consumers who cannot afford irregular and poor quality service substitute more expensive captive supply alternatives to minimize the negative consequences of power supply interruptions on their production activities and profitability. An estimated 20% of the investment into industrial projects is allocated to alternative sources of electricity supply[3].

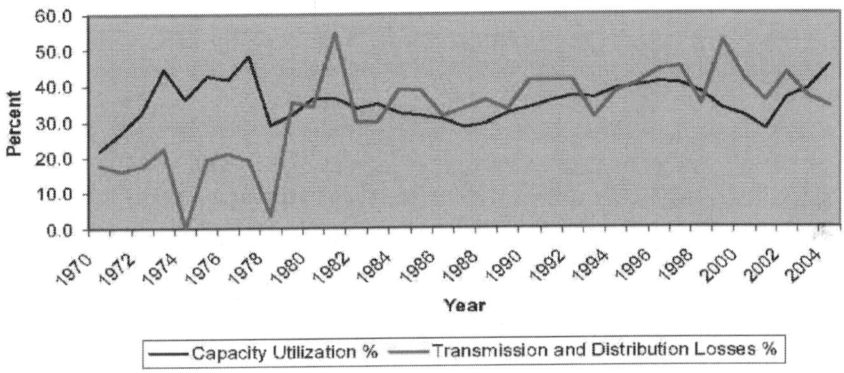

Figure 2 Indicators of the electricity crisis in Nigeria from 1970 to 2004. Adapted from Iwayemi[38].

In summary, the causal factors in Nigeria's energy crisis include the following:

Prevalence of a regime of price control.

Weak concern for cost recovery and lack of adequate economic incentives to induce the state-owned companies (NNPC and PHCN) to engage in efficient production and investment behavior. This seems apparent in the existence of large input and output subsidies.

Multiplicity of economic and noneconomic objectives without proper identification of the trade-offs among these different objectives. This is implicit in its pricing policies in both electricity and petroleum products markets.

Institutional and governance failures which induced gross distortions and inefficiency in production, investment choices and high costs of operation, low return on investment, and expensive delays along with cost overruns in the state energy enterprises.

1.5 Energy demand projection

There is an increasing demand for fuel energy due to the increase in economic development and civilization all over the world. Industry is one of the most important energy-consuming sectors in the world. According to Mitchel[41], energy is essential to our way of life. It provides us with comfort, transportation, and the ability to produce food and material goods. Historically, energy consumption has been directly related to the gross national product, which is a measure of the market value of the total national output of goods and services[42].

According to Sambo et al.[43], population is a major driver of energy demand, while its most important determinant is the level of economic activity and its structure measured by the total gross domestic product (GDP) alongside the various sectors and sub-sectors of the economy. Population projection of Nigeria was expected to grow from 115.22 million in 2000 to 281.81 million by 2030 at an average annual rate of 2.86% between 2000 and 2030.

Based on the models developed by the ECN, the country's energy demand was analyzed for the period from 2000 to 2030 with the use of the Model for the Analysis of the Energy Demand (MAED) and the Wien Automatic System Planning (WASP) package (Table5). It can be said that the energy demand of Nigeria will be approximately 2.5-, 3-, 3.5-, and 4.5-fold between the years 2000 and 2015 and approximately 8-, 13-, 17-, and 22.5-fold between the years 2000 and 2030 based on a 7% (reference), 10% (high growth), 11.5% (optimistic), and 13% (optimistic) GDP growth rate per annum, respectively. This increase in the energy demand is due to the high level of economic activities expected in Nigeria as measured by the total GDP.

Table 5 Total projected energy demand (MTOE)

Scenario	2000	2010	2015	2020	2025	2030
Reference (7%)	32.01	51.40	79.36	118.14	169.18	245.19
High growth (10%)	32.01	56.18	94.18	190.73	259.19	414.52
Optimistic (11.5%)	32.01	56.18	108.57	245.97	331.32	553.26
Optimistic (13%)	32.01	72.81	148.97	312.61	429.11	715.70

Adapted from ECN[44].

The trends of the projected energy demand are shown in Figure3. In 2005, the total energy demand based on a 10% GDP growth rate revealed that the household segment had the largest share of all the sectors. The sectorial energy demands in the 2030 plan period, however, showed the highest growth rates for the industrial, followed by the services, household, and transport sectors in that order (Table6). The electricity demand (extracted from the total energy demand) shows an increasing trend from the base year 2005 to 2030 in the four adopted growth scenarios, respectively, as shown in Figure4, indicating a high economic growth rate leading to a substantial increase in the electricity demand. The energy consumed over the years shows a decreasing trend with an increasing population, necessitating a corresponding increase in the energy output. Hence, the country's large energy efficiency potential needs to be exploited (Table7). In 2007, the total primary energy consumed was 11.4 million tons of oil equivalent (MTOE) with petroleum products having the largest share of 67.3% of the total consumption, amounting to a total average consumption of 78.7% between 2002 and 2007. This level of consumption was followed by that of hydropower at 23.9%, natural gas at 8.7%, and coal at 0.05% with their respective total average consumption standing at 16.08%, 5.17%, and 0.04% for the period from 2002 to 2007 as shown in Table8. Flaring adversely reduced the maximum contribution of natural gas to the total energy consumption mix in spite of its abundance in the country as most of the oil fields lack appropriate infrastructure for gas production. The general Niger Delta security issue (bunkering, sabotage, etc.) have also weakened most of the oil and gas projects[45].

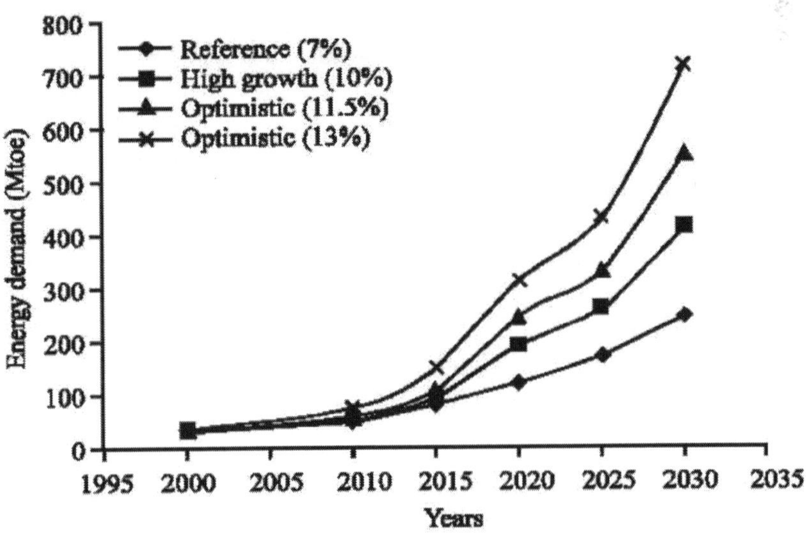

Figure 3 Graph showing the projected electricity demand between 2000 and 2030.

Table 6 Total energy demand based on a 10 % GDP growth rate (MTOE)

Item	2005	2010	2015	2020	2025	2030	Average growth rate (%)
Industry	8.08	12.59	26.03	39.47	92.34	145.21	16.2
Transport	11.70	13.48	16.59	19.70	26.53	33.36	4.7
Household	18.82	22.42	28.01	33.60	33.94	34.27	2.6
Services	6.43	8.38	12.14	15.89	26.95	38.00	8.7
Total	45.01	56.87	82.77	108.66	179.75	250.84	8.3

Adapted from ECN[46].

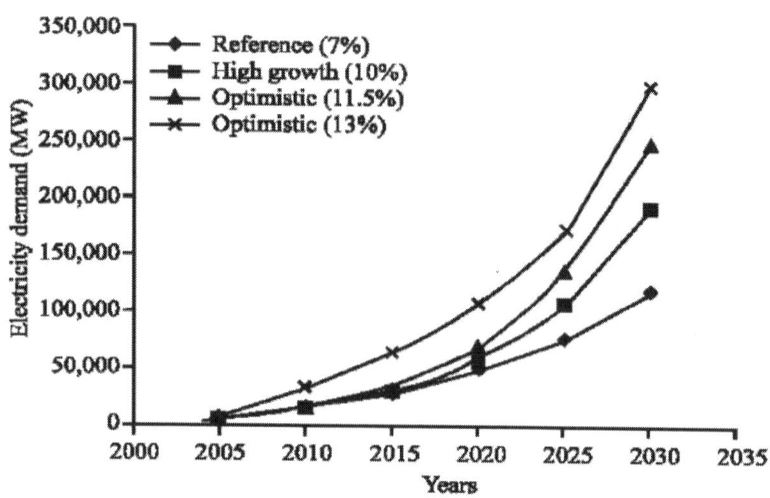

Figure 4 Graph showing the projected electricity demand between 2005 and 2030.

Table 7 *Per capita* primary energy consumption in Nigeria

Year	Energy consumed (MTOE)	Population (million)	*Per capita* energy consumption (TOE/capita)
2002	18.783	122.365	0.153
2003	19.106	126.153	0.151
2004	16.267	129.927	0.125
2005	17.707	133.702	0.132
2006	12.421	140.003	0.089
2007	11.387	144.203	0.097

Adapted from CBN and NBS[32, 47, 48].

Table 8 Commercial primary energy consumption by type (average percentage of total)

Type	2002	2003	2004	2005	2006	2007	Average
Coal	0.03	0.03	0.03	0.03	0.05	0.05	0.04
Hydro	11.93	14.20	17.39	12.04	17.03	23.90	16.08
Natural gas	2.84	1.9	4.54	5.5	7.52	8.73	5.17
Petroleum products	85.20	83.87	78.04	82.45	75.44	67.32	78.71

Adapted from CBN[32, 49].

Throughout the world, electricity is the most widely used and desirable form of energy. It is a basic requirement for economic development, national development, meeting the Millennium Development Goals (MDGs), and for an adequate standard of living. As a country's population grows and its economy expands, its demand for electrical energy multiplies. If this demand is not met adequately, a shortage in supply occurs. This shortage can assume crisis proportions and possibly affect achieving sustainable energy development.

The electric power capacity demand by projection in Nigeria would be approximately 3.5-fold between 2010 and 2020 and 7.5-fold between 2010 and 2030, respectively, at a growth rate of 7%, while the projected supply by fuel mix shows a similar trend with the demand at both growth rates of 7% and 13% (Table9). There is a wide disparity in the energy demand to the supply ratio in Nigeria both in the present and the future. This necessitates an urgent need for alternative energy sources and efficient energy usage in order to avert looming energy crises.

Table 9 Electric power capacity in Nigeria

	Electric power demand											
	2010				2020				2030			
	Demand (MW%)		Supply (MW%)		Demand (MW%)		Supply (MW%)		Demand (MW%)		Supply (MW %)	
Fuel type	7	13	7	13	7	13	7	13	7	13	7	13
Coal			0	0			6.515	16.913			15.815	63.896
Gas			13.555	31.935			37.733	78.717			85.585	192.895
Hydro			3.702	3.902			6.479	6.749			11.479	11.479
Nuclear			0	0			3.530	11.005			11.872	36.891
Small-hydro			40	208			140	1.000			701	2.353
Solar			5	30			34	750			302	4.610
Wind			0	500			1.471	3.791			5.369	15.567
Total	15.730	33.250	17.303	36.576	50.820	107.600	55.903	118.836	119.200	297.900	131.122	327.690

Supply by fuel mix and demand for 7% and 13% GDP growth. Adapted from ECN [44,46].

These projections for continued rapid energy growth imply some severe problems for the future resource depletion, energy degradation, associated environmental problems, fuel shortage, etc. Indeed, many of these problems are already happening; thus, energy conservation is concerned with ways to reduce energy demand, but yet achieve the same objective as before.

To achieve its objective of sustainable development, Nigeria needs to substantially increase the supply of modern affordable energy services to all its citizens while, at the same time, maintaining environmental integrity and social cohesion. In addition, a robust mix of energy sources (fossil and renewable), combined with an improved end-use efficiency, will almost certainly be required to meet the growing demand for energy services in the country. Technological development, decentralized non-grid networks, diversity of energy-supply systems, and affordable energy services are imperative to meeting the future demand.

1.6 The role of renewable energy technologies in sustainable development

Renewable energy has an important role to play in meeting the future energy needs in both rural and urban areas[50]. The development and utilization of renewable energy should be given a high priority, especially in the light of increased awareness of the adverse environmental impacts of fossil-based generation. The need for sustainable energy is rapidly increasing in the world. A widespread use of renewable energy is important for achieving sustainability in the energy sectors in both developing and industrialized countries.

Nigeria is blessed with a large amount of renewable natural resources (Table1), which, when fully developed and utilized, will lead to poverty reduction and sustainable development.

Renewable energy resources and technologies are a key component of sustainable development for the following primary reasons:

They generally cause less environmental impact than other energy sources. The implementation of renewable energy technologies will help to address the environmental concerns that emerged due to greenhouse gas emissions such as carbon dioxide (CO_2), oxides of nitrogen (NO_x), oxides of sulfur (SO_x), and particulate matters as a result of power generation from oil, natural gas, and coal. A variety of renewable energy resources provide a flexible array of options for their use.

They cannot be depleted. If used carefully in appropriate applications, renewable energy resources can provide a reliable and sustainable supply of energy almost indefinitely. In contrast, fossil fuel resources are diminished by extraction and consumption.

They favor system decentralization and local solutions that are somewhat independent of the national network, thus enhancing the flexibility of the system and providing economic benefits to small isolated populations.

To seize the opportunities presented by renewable energy resources in sustainable development, Nigeria needs to establish renewable energy markets and gradually develop experience with renewable energy technologies. The barriers and constraints to the diffusion of renewable energy should be overcome. A legal, administrative, and financing infrastructure should be established to facilitate planning and application of renewable energy projects. Government must play a useful role in promoting renewable energy technologies by initiating surveys and studies to establish their potential in both urban and rural areas.

Because renewable energies are constantly being replenished from natural resources, they have security of supply, unlike fossil fuels, which are negotiated on the international market and subject to international competition, sometimes even resulting in wars and shortages. They have important advantages, which could be stated as follows:

Their rate of use does not affect their availability in the future; thus, they are inexhaustible.

The resources are generally well distributed all over the world, even though wide spatial and temporal variations occur. Thus, all regions of the world have reasonable access to one or more forms of renewable energy supply.

They are clean and pollution-free and are therefore a sustainable natural form of energy.

They can be cheaply and continuously harvested and are therefore a sustainable source of energy.

Unlike the nuclear and fossil fuels plants which belong to big companies, governments, or state-owned enterprises, renewable energy can be set up in small units and is therefore suitable for community management and ownership. In this way, the returns from renewable energy projects can be kept in the community. In Nigeria, this has particular relevance since the electricity grid does not extend to remote areas, and it is prohibitively expensive to do so. This presents a unique opportunity to construct power plants closer to where they are actually needed. In this way, much needed income, skill transfer, and manufacturing opportunities for small businesses would be injected into rural communities.

1.7 Energy and sustainable development in Nigeria

Sustainable energy involves the provision of energy services in a sustainable manner, which in turn necessitates that energy services be provided for all people in ways that, now and in the future, are sufficient to provide the basic necessities, affordable, not detrimental to the environment, and acceptable to communities and people[51–53]. Linkages between sustainable energy and factors such as efficiency and economic growth have been investigated[54].

The energy sector plays a pivotal role in attempts to achieve sustainable development, balancing economic and social developments with environmental protection (encapsulated in the ‘strap line’ for the 2002 Johannesburg World Summit on Sustainable Development of ‘people, planet, and prosperity’). Energy is central to practically all aspects of sustainable development. Energy is central to the economy because it drives all economic activities. This characterization of energy directs our attention to its sources in nature, to activities that convert and reconvert this energy, and finally to activities that use the energy to produce goods and services and household consumption. Traditionally, energy is treated as an intermediate input in the production process. This treatment of energy's role understates its importance and contribution to development. All economic activities and processes require some form of energy. This effectively makes energy a critical primary factor of production. Given the state of technological advancement in the economy, capital and labor perform supporting roles in converting, directing, and amplifying energy to produce goods and services needed for growth and poverty reduction[3].

Energy services are essential ingredients of all three pillars of sustainable development - economic, social, and environmental. Economies that have replaced human and animal labor with more convenient and efficient sources of energy and technology are also the ones that have grown fastest. No country in modern times has succeeded in substantially reducing poverty without adequately increasing the provision and use of energy to make material progress[55]. Indeed, by not ensuring a minimum access to energy services for a broad segment of the population, economic development of developing countries such as Nigeria beyond the level of subsistence has proven to be a real challenge.

At the national level, energy propels economic development by serving as the launch pad for industrial growth and, via transport and communications, providing access to international markets and trade. Reliable, efficient, and competitively priced energy supplies also attract foreign investment - a very important factor in boosting economic growth in recent times. At the local level, energy facilitates economic development by improving productivity and enabling local income generation through improved agricultural development (irrigation, crop processing, storage, and transport to market) and through non-farm employment, including micro-enterprise development. As an indicator of local recognition of the importance of energy for businesses, Nigerian manufacturers, who were asked to rank the constraints on their firms' activities, identified power breakdowns, and voltage fluctuations as their top two problems[46]. Recent developments in Ghana's energy sector support this point[56].

Energy has also strong and important links to the environment. Many energy sources are drawn directly from the environment, requiring a sound management for these sources to be sustainable. Furthermore, energy use affects the environment. Emissions from fossil fuels, for example, reach beyond the local and national levels to affect the global environment and contribute to climate change. The poorest people often live in the most ecologically sensitive and vulnerable physical locations. These areas may be the most affected by the predictable effects of climate change such as an increased frequency of extreme events, for example floods, drought, rising sea levels, and melting ice caps. The risks facing poor people are often increased by the unsustainable use of biomass resources[3].

The connection between energy, the environment, and sustainable development is worth highlighting. Energy supply and use are related to climate change as well as such environmental concerns as air pollution, ozone depletion, forest destruction, and emissions of radioactive substances. These issues must be addressed if society is to develop while maintaining a healthy and clean environment. Ideally, a society seeking sustainable development should use only energy resources which have no environmental impact. However, since all energy resources lead to some environmental impact, an improved efficiency and environmental stewardship can help overcome many of the concerns regarding the limitations imposed on sustainable development by environmental emissions and their negative impacts[55].

Energy is directly linked to the broader concept of sustainability and affects most of civilization. That is particularly evident since energy resources drive much if not most of the world's economic activity, in virtually all economic sectors. Also, energy resources, whether carbon-based or renewable, are obtained from the environment, and wastes from energy processes (production, transport, storage, utilization) are typically released to the environment. Given the intimate ties between energy and the key components of sustainable development, the attainment of energy sustainability is being increasingly recognized as a critical aspect of achieving sustainable development[55].

Use of renewable natural resources, combined with efficient supply and use of fossil fuels with cleaner technologies, can help reduce the environmental effects of energy use and help Nigeria replacing the existing, inefficient fossil fuel technologies that pollute the environment.

As a complementary measure, careful management of energy resources is important to promote economic growth, protect ecosystems and provide sustainable natural resources.

Thus, energy sustainability is considered to involve the sustainable use of energy in the overall energy system. This system includes processes and technologies for the harvesting of energy sources, their conversion to useful energy forms, to provide energy services such as operating communications systems, lighting buildings, and cooking[57].

The reform of the energy sector is critical to sustainable development in Nigeria. This includes reviewing and reforming subsidies, establishing credible regulatory frameworks, developing policy environments through regulatory interventions, and creating market-based approaches such as emission trading[58]. Globally, countries are developing strategies and policies to enable a sustainable development of their energy resources, thus contributing to fuel economic and social developments, while reducing air pollution and greenhouse gas emissions.

The energy sector is very strategic to the development of the Nigerian economy. In addition to its macroeconomic importance, it has major roles to play in reducing poverty, improving productivity, and enhancing the general quality of life. If Nigeria is to take the path of sustainable energy, it is important to accurately and technically model the energy demand and supply scenarios and their impacts on the economy, resources, and society along with the environment, for both medium and long terms. From such analyses, we can derive information that is vital for policy construction and investment[59].

1.8 Energy efficiency and energy conservation in sustainable development

Energy efficiency means an improvement in practices and products that reduce the energy necessary to provide services. Energy efficiency products essentially help to do more work with less energy[60]. Energy efficiency is also defined as essentially using less energy to provide the same service[55]. In this sense, energy efficiency can also be thought of as a supply of resource - often considered an important, cost-effective supply option. Investment into energy efficiency can provide additional economic value by preserving the resource base (especially combined with pollution prevention technologies) and mitigating environmental problems.

Energy efficiency (EE) improvements have multiple advantages, such as the efficient exploitation of natural resources, the reduction in air pollution levels, and lower spending by the consumers on energy-related expenditure. Investments in EE result in long-term benefits, such as reduced energy consumption, local environmental enhancement, and overall economic development. Energy use has environmental impacts, regardless of the source or mechanism. For example, hydroelectric projects affect their local ecological systems and displace long-standing social systems. Fossil fuel power creates pollution in the extraction, transportation, and combustion of its raw materials. The long-term storage of waste products of the nuclear power industry is an issue to be resolved. Cost-effective energy efficiency is the ultimate multiple pollutant reduction strategy[61].

In Nigeria, a lot of energy is wasted because households, public and private offices, as well as industries use more energy than is actually necessary to fulfill their needs. One of the reasons is that they use outdated and inefficient equipment and production processes. Unwholesome practices also lead to energy wastage.

In Nigeria, the need for energy is exceeding its supply. In view of these circumstances, primary energy conservation, rationalization, and efficient use are immediate needs. Getting all the possible energy from the fuel into the working fluid is the goal of efficient equipment operations. This leads to a higher productivity and saves not only money, but also influences the safety and life of the equipment and reduces pollution[62]. Steps taken to minimize energy consumption, or to use the energy more effectively, are steps in the right direction to preserve the global environment. Energy conservation measures or recommendations are often referred to more positively as opportunities. Two primary criteria for applying energy conservation are that it is easy to implement and that its payback is brief. Ease of implementation and duration of payback period have been used to classify *Energy conservation opportunities* into three general categories for use: in maintenance and operation measures, in process improvement projects, and in large capital projects[61].

Energy conservation and energy efficiency are separate but related concepts. Energy efficiency is achieved when energy intensity in a specific product, process, or area of production or consumption is reduced without affecting output, consumption, or comfort levels. Promotion of energy efficiency will contribute to energy conservation and is therefore an integral part of energy conservation promotional policies[63].

Energy efficiency encompasses conserving a scarce resource; improving the technical efficiency of energy conversion, generation, transmission and end-use devices; substituting more expensive fuels with cheaper ones; and reducing or reversing the negative impact of energy production and consumption activities on the environment. Energy conservation is a tangible resource by itself that competes economically with contemporary energy supply options. In addition to this, it offers a practical means of achieving four goals that should be of high priority in any nation that desires quick and sustainable economic growth and development. These are economic competitiveness, utilization of scarce capital for development, environmental quality, and energy security. It enhances the international competitiveness of the industries in the world markets by reducing the cost of production. It optimizes the use of capital resources by directing lesser amounts of money in conservation investment as compared with capital-intensive energy supply options. It protects the environment in the short run by reducing pollution and in the long run by reducing the scope of global climate change. It strengthens the security of supply through a lesser demand and a lesser dependence on petroleum product imports. No energy supply option may be able to provide all these benefits. Energy conservation is a decentralized issue and is largely dependent on individual, distinct decisions of energy supply, which are highly centralized. The housewife, the car driver, the housing developer, the house owner, the boiler operator in industry, and every other individual who consumes energy in some form or another are required to participate in energy-saving measures. It calls for a collective endeavor and is dependent upon the actions of people in diverse fields although the people involved may not be sufficiently informed or motivated to conserve energy[64].

1.9 Renewable energy and energy efficiency as climate change mitigation strategies

The Inter-government Panel on Climate Change (IPCC), a body set up in 1988 by the World Meteorological Organization and the United Nations Environmental Programme to provide authoritative information about the climate change phenomenon, asserts that the warming of the last 100 years was unusual and unlikely to be natural in origin[58]. The IPCC has attributed the warming of at least the second half of the century to an increase in the emission of greenhouse gases into the atmosphere. Human activity is largely responsible for the emission of these gases into the atmosphere: CO_2 is produced by the burning of fossil fuels (coals, oil, gas) as well as by land-use activities such as deforestation; methane is produced by cattle, rice agriculture, fossil fuel use, and landfills; and nitrous oxide is produced by the chemical industry, cattle feed lots, and agricultural soils. As humans have increased their levels of production and consumption, greenhouse gas emissions have also increased; since 1750, at the time of the industrial revolution, CO_2 emission has increased by 31%, methane by 15%, and nitrous oxide by 17%. Moreover, the emissions of these gases continue to rise steadily[65].

The Clean Development Mechanism (CDM) was integrated to the Kyoto Protocol as the United Nations Framework Convention on Climate Change[66]. CDM projects allow investment by entities from industrialized countries into projects in developing countries. In return for this investment, carbon credits (in this case, certified emission reductions) are received by the investor in the industrialized country. This enables the industrialized country to meet its emission reduction targets given by the Kyoto Protocol more cost-effectively, while promoting sustainable development in developing countries. CDM projects may also be unilateral, i.e., they take place in the developing country without a project partner from an industrialized nation.

Investment into clean energy facilities is recognized as the best way to increase the participation of Nigerian proponents in the CDM process and hence the global carbon market. Clean energy investment is defined as follows: *investment into an energy supply and utilization system that provides the required energy with minimal negative environmental and social consequences*[67]. Investment into clean energy systems can also be viewed as an investment into energy sources and technologies that are significantly less environmentally damaging than in the*status quo* case. Investment into clean energy systems provides the most effective and optimally efficient path to an increased CDM participation in Nigeria and hence an effective participation in the global carbon market.

The salient characteristics of clean energy investment are as follows:

The resulting system results in little or no emissions of obnoxious gases and particulates;

The clean energy technologies have a carbon footprint that is much lower than the baseline emission scenario;

The technology is accessible, and the required investment is available for adoption in developing countries like Nigeria;

The implementation of the clean energy technology will contribute to sustainability.

Energy efficiency and renewable energy technologies are prominent in most sustainable development programs, for example, the Agenda 21[68]. According to the Intergovernmental Panel on Climate Change (IPCC), the second assessment report, the stabilization of atmospheric greenhouse gas concentrations at levels that will prevent serious interference with the climate system can only be achieved by dramatically increasing the implementation of renewable energy. In one IPCC scenario, in which greenhouse gases are stabilized by the year 2050, the share of renewable energy in the global energy balance must increase tenfold from the current level. In developing countries, the required increase is even more dramatic, estimated at 20-fold between 1990 and 2050. Further improvements in energy efficiency and energy conservation can reduce emissions in the shorter term, thus 'buying time' for the required changes in energy production[69].

Nigeria is one of the highest emitters of greenhouse gases in Africa. The practice of flaring gas by the oil companies operating in Nigeria has been a major means through which greenhouse gases are released into the atmosphere. Carbon dioxide emissions in this area are among the highest in the world[70]. Some 45.8 billion kW of heat are discharged into the atmosphere of the Niger Delta from flaring 1.8 billion ft^3 of gas every day[58]. Gas flaring has raised temperatures and rendered large areas uninhabitable. Between 1970 and 1986, a total of about 125.5 million m^3 of gas was produced in the Niger Delta region, about 102.3 (81.7 %) million m^3 were flared, while only 2.6 million m^3 were used as fuel by oil-producing companies and about 14.6 million m^3 were sold to other consumers[71, 72]. The use of renewable energy sources will reduce the over dependence on the burning of fossil fuel. Moreover, instead of flaring gas in Nigeria, the gases can be converted to methanol and used as a fuel for both domestic and industrial use. With good energy efficiency practices and products, the burning of fossil fuel for energy will be greatly minimized.

2. CONCLUSIONS

From the energy outlook of Nigeria, it is very clear that the energy demand is very high and is increasing geometrically while the supply remains inadequate, insecure, and irregular and is decreasing with time; the mix has hitherto been dominated by fossil resources which are fast being depleted apart from being environmentally non-friendly. The energy supply mix must thus be diversified through installing an appropriate infrastructure and creating full awareness to promote and develop the abundant renewable energy resources present in the country as well as to enhance the security of supply.

There is clear evidence that Nigeria is blessed with abundant resources of fossil fuels as well as renewable energy resources. The major challenge is an inefficient usage of energy in the country. As a result, there is an urgent need to encourage the evolvement of an energy mix that will emphasize the conservation of petroleum resources in such a manner enabling their continued exportation for foreign earnings for as many years as possible.

The opportunities for conserving energy in our various sectors - office building and residential areas, manufacturing industries, transportation, electricity generation and distribution, and electricity equipment and appliances - were presented in this work. The various areas where savings in energy can be made have also been identified. Several guidelines and measures have been suggested to conserve energy in these areas, and if the guidelines and measures are strictly adhered to, then substantive savings in energy will be carried out.

In this study, four economic growth scenarios were dealt with in the review of the energy requirements. These are the reference scenarios of a 7% total GDP growth rate that will ensure the MDGs of reducing poverty by 50% of the 2,000 value by 2015. The high growth scenario of a 10% GDP growth rate in the attempt to eradicate poverty by 2030 and the optimistic scenarios of 11.5% and 13% GDP growth rates that will further increase the rate of economic development.

In order to ensure the sustainability of energy supply and subsequently the sustainable economic development of the country, the government has to intensify the further implementation of renewable energy and energy efficiency programs. As observed in quite a number of successful countries promoting renewable energy, such as Germany, Denmark, and Japan, a strong and long-term commitment from the government is crucial in implementing any kind of policies which will lead to the development of renewable energies, in particular, and a sustainable development, in general.

3. RECOMMENDATIONS

In this study, it is established that renewable energy and energy efficiency are two components that should go together to achieve sustainable development in Nigeria. The need to conserve the present energy generated in the country using energy-efficient products and the appropriate practices is essential for sustainable development. Therefore, it is recommended that the country should do the following:

Develop policies on energy efficiency and integrate them into the current energy policies. A comprehensive and coherent energy policy is essential in guiding the citizens towards an efficient usage of its energy resources.

Promote energy-efficient products and appropriate practices at the side of the end users and energy generation.

Create awareness on renewable energy and energy efficiency.

Establish an agency to promote the use of energy-efficient products and ensure the appropriate practices.

Develop and imbibe energy efficiency technologies.

Carry out a resource survey and assessment to determine the total renewable energy potential in the country as well as identify the local conditions and local priorities in various ecological zones.

Establish a testing and standards laboratory for renewable energy technologies similar to that in South Africa.

Take advantage of global partnerships, such as the Residential Energy Efficiency Project initiative of UK, to assist the country in a creative integration of renewable energy systems.

Establish a renewable energy funding/financing agency such as India's Indian Renewable Energy Agency.

Develop appropriate drivers for the implementation of energy efficiency policies.

Clean energy facilities should be embraced in the different sectors of the Nigerian economy.

In the following, a partial list of potential clean energy opportunities in Nigeria is presented:

More efficient passive and full usage of solar technologies in the residential, commercial, and industrial sectors.

Biogas from wastes as a source of cooking fuel in homes.

Use of energy-efficient lighting.

Implementation of renewable biomass as a fuel in highly efficient cook stoves.

Efficient production of charcoal as a fuel in homes and *small and medium enterprises*.

Use of biofuels in efficient cooking stoves and lamps in homes.

Energy-efficient lighting.

Use of compressed natural gas (CNG) as a transport fuel.

Use of biofuels as a transport fuel.

Introduction of a bus rapid transit system to other cities and expansion of the Lagos system.

Shift from high carbon intensive fuels to natural gas for energy generation in industries.

Development of a CNG infrastructure to distribute natural gas to industries located at sites remote from the existing pipelines.

Implementation of combined heat and power (CHP) facilities in industries.

Implementation of energy efficiency improvements in manufacturing industries.

Implementation of CHP facilities in commercial facilities.

Use of solar and wind energy for irrigation water pumping and farm electricity supply.

Utilization of agricultural residues for electricity generation.

Generation of biogas from wastes produced by the livestock and animal husbandry.

In addition to these, the existing research and development centers and technology development institutions should be adequately strengthened to support the shift towards an increased use of renewable energy. Human resource development, critical knowledge, and know-how transfer should be the focus for project development, project management, monitoring, and evaluation. The preparation of standards and codes of practices, maintenance manuals, life cycle costing, and cost-benefit analysis tools should be undertaken on urgent priority.

COMPETING INTERESTS

The author declares that there are no competing interests.

REFERENCES

1 Ramchandra P, Boucar D: *Green Energy and Technology*. Springer, London Dordrecht Heidelberg New York; 2011.

2 Rai GD: *Non- Conventional Energy Sources*. Khanna Publishers, Delhi; 2004.

3 Nnaji C, *et al.*: *CIA World Factbook*. Edited by: Nnaji CE, Uzoma CC. Nigeria; 2010. http://www.cia.gov/library/publications/the-world-factbook/geos/ni.html

4 Council for Renewable Energy, Nigeria (CREN): *Nigeria Electricity Crunch*. 2009. available atwww.renewablenigeria.org

5 CBN: *Central Bank of Nigeria Annual Reports and Statement of Account*. 1985.

6 Department of Petroleum Resources (DPR) Nigeria, Nigeria; 2007.http://www.DPR.gov.ng available at

7 Onyebuchi EI: **Alternative energy strategies for the developing world's domestic use: A case study of Nigerian household's final use patterns and preferences.** *The Energy Journal* 1989,**10**(3):121–138.

8 Chineke TC, Igwiro EC: **Urban and rural electrification: enhancing the energy sector in Nigeria using photovoltaic technology.***African Journal Science and Tech* 2008,**9**(1):102–108.

9 Energy Commission of Nigeria (ECN): *Renewable Energy Master Plan*. 2005.

10 Adekoya LO, Adewale AA: **Wind energy potential of Nigeria.** *Renewable Energy* 1992,**2**(1):35–39. 10.1016/0960-1481(92) 90057-A

11 Fagbenle RO, Karayiannis TG: **On the wind energy resources of Nigeria.** *International Journal of Energy research* 1994,**18**(5):493–508. 10.1002/er.4440180502

12 Ngala GM, Alkali B, Aji MA: **Viability of wind energy as a power generation source in Maiduguri, Borno state, Nigeria.** *Renewable energy* 2007,**32**(13):2242–2246. 10.1016/j.renene.2006.12.016

13 Akinbami JFK: *Renewable Energy Resources and Technologies in Nigeria: Present Situation, Future Prospects and Policy Framework'. Mitigation and Adaptation Strategies for Global Change 6:155–181.* Kluwer Academic Publishers, Netherlands; 2001.

14 Akinbami JFK, Ilori MO, Oyebisi TO, Akinwumi IO, Adeoti O: **Biogas energy use in Nigeria: Current status, future prospects and policy implications.** *Renewable and Sustainable Energy Review* 2001, **5:** 97–112. 10.1016/S1364-0321(00)00005-8

15 Sambo AS: **Matching Electricity Supply with Demand in Nigeria.** *International Association of Energy Economics* 2008, **4:** 32–36.

16 Ighodaro CAU: **Co-Integration And Causality Relationship Between Energy Consumption And Economic Growth: Further Empirical Evidence For Nigeria.** *Journal of Business Economics and Management* 2010,**11**(1):97–111. 10.3846/jbem.2010.05

17 Okafor ECN, Joe-Uzuegbu CKA: **Challenges To Development Of Renewable Energy For Electric Power Sector In Nigeria.***International Journal Of Academic Research* 2010,**2**(2):211–216.

18 Energy Commission of Nigeria (ECN): *Draft National Energy Masterplan*. 2007.

19 *Report of the Inter-Ministerial Committee on Combating Deforestation and Desertification*. August 2000.

20 Famuyide OO, Anamayi SE, Usman JM: **Energy Resources' Pricing Policy And Its Implications On Forestry And Environmental Policy Implementation In Nigeria.** *Continental J Sustainable Development* 2011, **2:** 1–7.

21 Sambo AS: **Strategic Developments in Renewable Energy in Nigeria.** *International Association of Energy Economics* 2009, **4:** 15–19.

22 Energy Commission of Nigeria (ECN): *National Energy Policy*. Federal Republic of Nigeria, Abuja; 2003.

23 Abiodun R: *Fuel price Hike Spells Doom for Nigeria's Forest*. 2003. htt:www.islamonline.net/English/index.shtml

24 Williams CE: *'Reaching the African Female Farmers with Innovative Extension Approaches: Success and Challenges for the Future'*. Paper Presented To The International Workshop on Women Agricultural Intensification and Household Food Security At University of Cape Coast, Ghana; 1998. 25th-28th June 25th-28th June

25 Karekezi R: *Renewable Energy Technologies in Africa, Zed Books Limited with African Energy Policy Research Network (AFREPREN) and Stockholm Environment Institute (SEI)*. 1997.

26 Central Bank of Nigeria (CBN): *'The changing structure of the Nigerian economy and implications for development'*. Research Department, Central Bank of Nigeria; Realm Communications Ltd, Lagos; 2000. August August

27 Famuyide OO, Adu AO, Ojo MO: **Socio-Economic Impacts of Deforestation in the Sudano- Sahelian Belt of Nigeria.** *Journal of Forestry Research and Management* 2004,**1**(1&2):94–106.

28 Fagbenle RO, Adenikinju A, Ibitoye FI, Yusuf AO, Alayande O: *Draft Final Report on Nigeria's Electricity Sector Executive Report*. 2006.

29 PHCN: *Generation and Transmission Grid Operations*. Annual Technical Report National Control Center (NCC), Osogbo; 2008.

30 PHCN: *Generation and Transmission Grid Operations*. National Control Center (NCC), Osogbo; 2009.

31 Imo EE: **'Challenges of Hydropower Development in Nigeria'.** *Hydrovision* 2008.

32 CBN (Central Bank of Nigeria) 2007. Vol. 18th edn., , December

33 Emeka EE: **'Causality Analysis of Nigerian Electricity Consumption and Economic Growth'.** *Journal of Economics and Engineering*2010, **4:** 80–85. ISSN: 2078–0346 ISSN: 2078–0346

34 Odularu GO, Okonkwo C: **'Does Energy Consumption Contribute To Economic Performance? Empirical Evidence From Nigeria', Journal of Economics and Business Vol.** *XII, No* 2009, **2:** 43–47.

35 CBN: *Central Bank of Nigeria Statistical Bulletin.* CBN Press, Abuja; 2009.

36 UNIDO (United Nations Industrial Development Organization): *UNIDO (United Nations Industrial Development Organization).* UNIDO (United Nations Industrial Development Organization), UNIDO (United Nations Industrial Development Organization); 2001.

37 Ajayi OO, Ajanaku KO: **Nigeria's Energy Challenge and Power Development: The Way Forward.** *Bulletin of Science Association of Nigeria* 2007, **28:** 1–3.

38 Iwayemi A: *Nigeria's Dual Energy Problems: Policy Issues and Challenges. International Association for Energy Economics.* 2008, 17–21. Fourth Quarters Fourth Quarters

39 Ibitoye F, Adenikinju A: **(2007) Future Demand for Electricity in Nigeria.** *Applied Energy* 2007, **84:** 492–504. 10.1016/j. apenergy. 2006.09.011

40 Adenikinju A: *' Analysis of the cost of infrastructure failures in a developing economy the case of electricity sector in Nigeria'.* African Economic Research Consortium AERC Research Paper 148, Nairobi; 2005. February 2005 February 2005

41 Mitchel JW: *Energy engineering.* John Wiley and Sons, New York; 1983.

42 Adeyemo SB: **'Energy potentials of organic wastes'.** *Proceedings of the first national conference* 2001, 55–61. ISBN 978–35533–0-5 ISBN 978-35533-0-5

43 Sambo AS, Iloeje OCJ, Ojosu OJ, Olayande S, Yusuf AO: *Nigeria's Experience on the Application of IAEA"s Energy Models (MAED & WASP) for National Energy Planning, Paper presented during the Training Meeting /Workshop on Exchange of Experience in Using IAEA's Energy Models and Assessment of Further Training Needs.* held at the Korea Atomic Energy Research Institute, Daejon, Republic of Korea; 2006. pp 24–28, April pp 24–28, April

44 Energy Commission of Nigeria (ECN): *National Energy Policy.* Federal Republic of Nigeria, Abuja; 2006.

45 Ohunakin OS: **Energy utilization and renewable energy sources in Nigeria.** *Journal of Engineering and Applied Sciences*2010,**5**(2):171–7.

46 Energy Commission of Nigeria (ECN): *National Energy Policy.* Federal Republic of Nigeria, Abuja; 2008.

47 CBN: *Central Bank of Nigeria Statistical Bulletin.* CBN Press, Abuja; 2006.

48 National Bureau of Statistics (NBS): *National Core Welfare Indicator Questionnaire (CWIQ) Survey.* National Bureau of Statistics (NBS), Abuja; 2007. p 2, Summary Sheet p 2, Summary Sheet

49 CBN: *Central Bank of Nigeria Statistical Bulletin.* CBN Press, Abuja; 2005.

50 Hui SCM: **From renewable energy to sustainability: the challenge for Honk Kong'.** *Hong Kong Institution of Engineers* 1997, 351–358.

51 Lior N: **Energy resources and use: The present situation and possible paths to the future.** *Energy* 2008, **33:** 842–857. 10.1016/j.energy.2007.09.009

52 Haberl H: **The global socioeconomic energetic metabolism as a sustainability problem.** *Energy* 2006, **31:** 87–99. 10.1016/j.energy.2004.04.045

53 Rosen MA: **Energy efficiency and sustainable development.** *Int J Global Energy Issues* 2002, **17:** 23–34.

54 Ayres RU, Turton H, Casten T: **Energy efficiency, sustainability and economic growth.** *Energy* 2007, **32:** 634–648. 10.1016/j.energy.2006.06.005

55 Rosen MA: **Energy Sustainability: A Pragmatic Approach and illustrations.** *Sustainability* 2009, **1:** 55–80. 10.3390/su1010055

56 Ashong SN: *Macroeconomic Framework for Poverty Reduction Within the Context of Debt Relief:The Case of Ghana.* Paper Presented at WIDER Development Conference on Debt Relief Helsinki, Finland; 2001.

57 Hammond GP: **Alternative energy strategies for the United Kingdom revisited; market competition and sustainability.***Technological Forecasting and Social Change* 1998, **59:** 131–151. 10.1016/S0040-1625(97)00151-0

58 Sims REH, Schock RN, Adegbululgbe A, Fenhann J, Konstantinaviciute I, Moomaw W, Uyigue E: *CREDC Conference on Promoting Renewable Energy and Energy Efficiency in Nigeria.* held at University of Calabar Hotel and Conference Centre; 2007. 21st November 21st November

59 Winkler H (Ed): *Energy policies for sustainable development in South Africa Options for the future.* Energy Research Centre University of

Cape Town, Private Bag Rondebosch 7701 South Africa; 2006. Website:http://www.erc.uct.ac.za Website:

60 SECCP: *'Getting to grips with sustainable energy'*. 2002.

61 Adeyemo SB, Odukwe AO: **'Energy Conservation as a Viable Pathway towards Energy Stability.** *Journal of Engineering and Applied Sciences* 2008,**3**(3):233–238.

62 Habib MA, Said SAM, Igbal MO, El-Mahallawy FM, Mahdi EA: *Energy Conservation and Early Failure Prediction in Boilers and Industrial Furnaces*. Symposium on Management of Energy Consumption in Industry, Chamber of Commerce, Dammam, Kingdom of Saudi Arabia, October; 1999.

63 Dubin FS, Long CG: *'Energy conservation standards for Building design, construction and operation*. McGraw- Hill Book Company; 1980.

64 Etiosa U (Ed): *Energy Efficiency Survey in Nigeria: A Guide for Developing Policy and Legislation*. International, Rivers; 2009:1–37.

65 Uyigue E: **'Renewable energy and energy efficiency and sustainable development in Nigeria'.** *CREDC Conference on Promoting Renewable Energy and Energy Efficiency in Nigeria* 2007.

66 Sarah La P: *'Climate Change and Poverty'*. A Publication of Tearfund; 2002.

67 Winkler H, Van Es D: **Energy Efficiency and CDM in South Africa: Constraints and Opportunities.** *Journal of Energy in South Africa*2007, **18:** 29–37.

68 Dayo FB: *Clean Energy Investment in Nigeria The domestic context*. International Institute for Sustainable Development (IISD); 2008.

69 The United Nations (UN): *Report on the National Fuel wood Substitution Programme*. United Nations Publications, New York, Energy Statistics Yearbook; 1993.

70 Martinot E, McDom O: *Promoting Energy Efficiency and Renewable Energy GEF Climate Change Projects and Impacts; Global Environmental Facility*. Washington, DC; 2002.

71 Iyayi F: *'An integrated approach to development in the Niger Delta'*. A paper prepared for the Centre for Democracy and Development (CDD); 2004.

72 Awosika LF: **Impacts of global climate change and sea level rise on coastal resources and energy development in Nigeria.** In*Global Climate Change:Impact on Energy Development*. Edited by: Umolu JC. DAMTECH Nigeria Limited, Nigeria; 1995.

CHAPTER 5

Potential of wind power projects under the Clean Development Mechanism in India

Pallav Purohit[1] and Axel Michaelowa[2]

[1]Research Programme on International Climate Policy, Hamburg Institute of International Economics (HWWI)

[2]Political Economy and Development, Institute of Political Science, University of Zurich

ABSTRACT

Background

So far, the cumulative installed capacity of wind power projects in India is far below their gross potential (≤ 15%) despite very high level of policy support, tax benefits, long term financing schemes etc., for more than 10 years etc. One of the major barriers is the high costs of investments in these systems. The Clean Development Mechanism (CDM) of the Kyoto Protocol provides industrialized countries with an incentive to invest in emission reduction projects in developing countries to achieve a reduction in CO_2 emissions at lowest cost that also promotes sustainable development in the host country. Wind power projects could be of interest under the CDM because they directly displace greenhouse gas emissions while contributing to sustainable rural development, if developed correctly.

Results

Our estimates indicate that there is a vast theoretical potential of CO_2 mitigation by the use of wind energy in India. The annual potential Certified Emissions Reductions (CERs) of wind power projects in India could theoretically reach 86 million. Under more realistic assumptions about diffusion of wind power projects based on past experiences with the government-run programmes, annual CER volumes by 2012 could reach 41 to 67 million and 78 to 83 million by 2020.

Conclusion

The projections based on the past diffusion trend indicate that in India, even with highly favorable assumptions, the dissemination of wind power projects is not likely to reach its maximum estimated potential in another 15 years. CDM could help to achieve the maximum utilization potential more rapidly as compared to the current diffusion trend if supportive policies are introduced.

Background: Indian Energy Demand, Wind Energy Technology, Technical Potential of Wind Energy and its Support in India

The global energy demand is expected to grow at a staggering rate in the next 30 years. The International Energy Agency [1] predicts that the world's energy needs will be almost 60% higher in 2030 than they are now. Two-thirds of this increase will arise in China, India and other rapidly developing economies, which will account for almost half the energy consumption by 2030. Sharp increases in world energy demand will trigger important investments in generating capacity and grid infrastructure. According to the IEA, the global power sector will need to build some 4,800 GW of new capacity between now and 2030.

In the 11th Five Year Plan, the Government of India aims to achieve a GDP growth rate of 10% and maintain an average growth of about 8% in the next 15 years [2]. According to Indian government officials, the growth of Indian economy is highly dependent on the growth on its energy consumption [3]. The 2006 capacity of power plants in India was 124 GW, of which 66% thermal, 25% hydro, 3% nuclear and 5% new renewables [4]. At the same time, Chinese power capacity reached over 600 GW [5], showing India's backlog. Wind energy is an alternative clean energy source and has been the world's fastest growing renewable energy source growing at a rate of 28% in the last decade [6]. Wind power has the advantage of being harnessed on a local basis for application in rural and remote areas [7]. Global wind power capacity reached 74 GW at the end of 2006 [8], 13 countries had more than 1 GW installed. Figure 1 presents the regional distribution of the global installed wind power capacity [8].

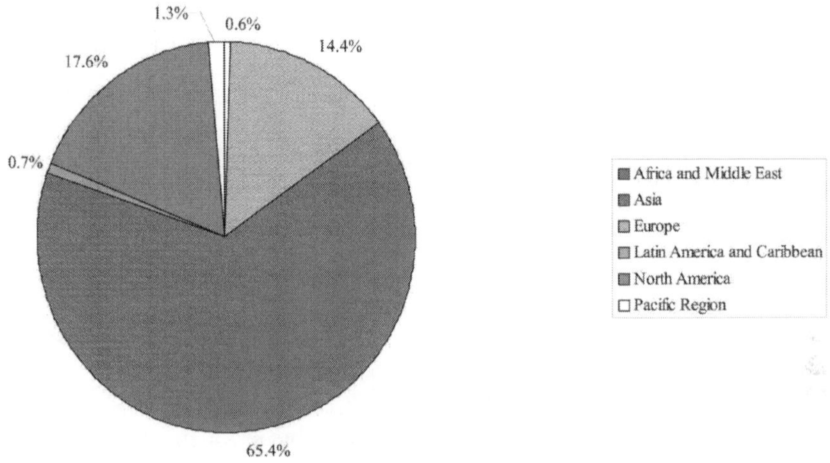

Figure 1 Regional distribution of the global installed wind power capacity (Source: [8]).

The impetus behind wind power expansion has come increasingly from the urgent need to combat global climate change. Most countries now accept that greenhouse gas (GHG) emissions must be drastically slashed in order to avoid environmental catastrophe. Wind energy offers both a power source that completely avoids the emission of carbon dioxide, the main GHG, but also produces none of the other pollutants associated with either fossil fuel or nuclear generation [9]. Wind power can deliver industrial scale on-grid capacity. Starting from the 1997 Kyoto Protocol, a series of GHG reduction targets has cascaded down to a regional and national level. These in turn have been translated into targets for increasing the proportion of renewable energy, including wind. In order to achieve these targets, countries in both Europe and elsewhere have adopted a variety of market support mechanisms [10–15]. These range from premium payments per unit of output to more complex mechanisms based on an obligation on power suppliers to source a rising percentage of their supply from renewables. As the market has grown, wind power has shown a dramatic fall in cost [16]. The production cost of a kilowatt-hour of wind power is one fifth of what it was 20 years ago. In the best locations, wind is already competitive with new coal-fired plants. Individual wind turbines have also increased in capacity, with the standard commercial machines reaching 2.5 MW and prototypes for offshore plants even 5 MW.

The successful wind energy business has attracted the serious attention of the banking and investment market, with new players such as oil companies entering the market. Hays and Attwood [17] concludes that Asia is playing an increasingly important role in the global wind industry as the region prepares to invest over $12 billion in wind power generation capacity in the second half of this decade. In India, wind power already occupies a prominent position with regard to installed capacity – reaching 6.2 GW by the end of 2006. In 2006 alone, an aggregate capacity of 1.8 GW has been added [8]. Thus, India is the fourth largest wind market in the world [18]. However, the total installed capacity of wind power projects still remains far below from their respective potential (i. e. <15%). One of the barriers to the large-scale dissemination of wind power projects in India is the high upfront cost of these systems [19]. Other barriers to wind power projects are low plant load factors, unstable policies of the state governments and poor institutional framework.

Wind has considerable amount of kinetic energy when blowing at high speeds [20]. This kinetic energy when passing through the blades of the wind turbines is converted into mechanical energy and rotates the wind blades [21] and the connected generator, thereby producing electricity. A wind turbine primarily consists of a main tower, blades, nacelle, hub, main shaft, gearbox, bearing and housing, brake, and generator [22]. The main tower is 50–100 m high. Generally, three blades made up of Fiber Reinforced Polyester are mounted on the hub, while in the nacelle the major parts are housed. Under normal operating conditions the nacelle would be facing the upstream wind direction [20]. The hub connects the gearbox and the blades. Solid high carbon steel bars or cylinders are used as main shaft. The gearbox is used to increase the speed ratio so that the rotor speed is increased to the rated generator speed [21]; it is the most critical component and needs regular maintenance. Oil cooling is employed to control the heating of the gearbox. Gearboxes are mounted over dampers to minimize vibration. Failure of gearbox may put the plant out of operation for an entire season as spares are often not available. Thus, new gearless configurations have become attractive for wind plant operators. Modern turbines fall into two basic groups: horizontal axis turbines and vertical axis turbines as shown in Figure 2[23]. Horizontal axis turbines resemble airplane propellers, with two to three rotor blades fixed at the front of the tower and facing into the wind. This is the most common design found today, making up most of the large utility-scale turbines on the global market. Vertical axis turbines resemble a large eggbeater with rotor blades attached vertically at the top and near the bottom of the tower and bulging out in the middle.

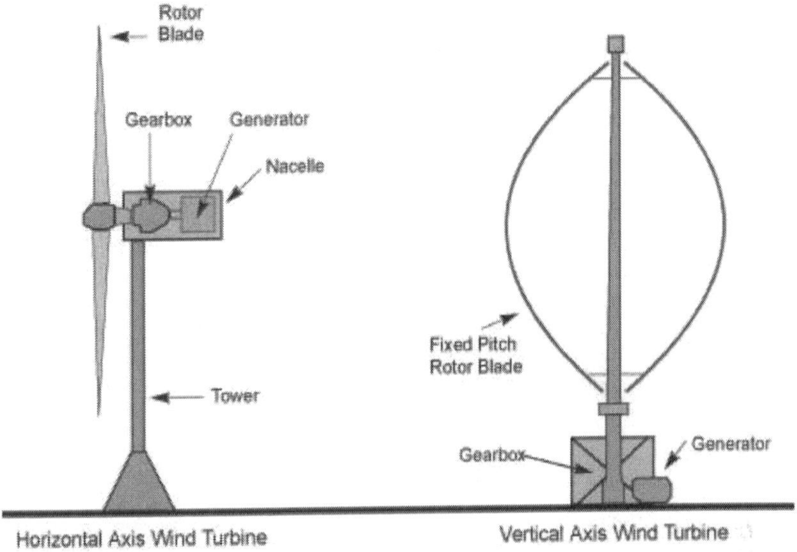

Figure 2 Schematic of the horizontal and vertical axis wind turbine (Source: [23]).

The most dramatic improvement has been in the increasing size and performance of wind turbines. From machines of just 25 kW twenty years ago, the commercial size range sold today is typically from 600 up to 2,500 kW, with 80 m diameter rotors placed on 70–100 m high towers. In 2003, the German company Enercon erected the first prototype of a 4.5 MW turbine with a rotor diameter of 112 m. Wind turbines have a design lifetime of 20–25 years, with their operation and maintenance costs typically about 3–5% of the cost of the turbine. For the share of different wind turbine types in India see Table 1.

Table 1 Manufacturers-wise wind electric generators installed in India (as on 31.03.2006)

S. No.	Manufacturer	Rating (kW)	Details as on 31st March 2006	
			Numbers	Capacity (in MW)
1	ABAN – Kenetech	410	231	94.71

S. No.	Manufacturer	Rating (kW)	Details as on 31st March 2006	
2	AMTL – Wind World	220	2	0.44
		250	328	82.00
		500	3	1.50
3	BHEL	55	16	0.88
		200	17	3.40
4	BHEL Nordex	200	79	15.80
		250	184	46.00
5	C-WEL	250	57	14.25
		600	2	1.20
6	Danish Windpower	150	12	1.80
7	Das Lagerwey	80	9	0.72
		250	284	71.00
8	Elecon	200	1	0.20
		300	51	15.30
		600	5	3.0
9	Enercon	230	451	103.73
		330	38	12.54
		600	681	408.60

S. No.	Manufacturer	Rating (kW)	Details as on 31st March 2006	
		800	435	348
10	GE Wind Energy	1500	12	18
11	Himalaya	140	4	0.56
		200	24	4.80
12	JMP-Ecotecnia	225	10	2.25
13	Kirloskar – WEG	400	8	3.20
14	Micon (Pearl)	90	99	8.91
15	Mitsubishi	315	6	1.89
16	Nedwind-Windia	250	4	1.00
		500	20	10.00
		550	35	19.25
17	NEG Micon	750	674	505.5
		950	54	51.30
		1650	137	226.05
18	NEPC India	225	957	215.325
		250	16	4.0
		400	7	2.80
		750	12	9.0

S. No.	Manufacturer	Rating (kW)	Details as on 31st March 2006	
19	NEPC-Micon	55	14	0.77
		110	2	0.22
		200	50	10.00
		225	589	132.53
		250	528	132.00
		400	121	48.40
		600	2	1.20
20	Pegasus	250	9	2.25
21	Pioneer Asia	850	35	29.75
22	Pioneer Wincon	110	10	1.10
		250	260	65.00
		755	1	0.76
23	REPL-Bonus	55	22	1.21
		100	1	0.10
		320	60	19.20
24	RES-Adavanced Wind Turbine	250	80	20.00
25	Sangeeth – Carter	300	25	7.50
26	Suzlon	270	2	0.54

S. No.	Manufacturer	Rating (kW)	Details as on 31st March 2006	
		350	836	292.60
		600	15	9.0
		1000	81	81.00
		1250	1255	1568.75
		2000	1	2.00
27	Tacke	250	4	1.00
		600	21	12.60
		750	1	0.75
28	Textool-Nordtank	300	65	19.50
		550	5	2.75
29	TTG/Shriram EPC	250	230	57.50
30	Vestas – RRB	55	31	1.71
		90	21	1.89
		100	5	0.50
		200	56	11.20
		225	735	165.375
		500	562	281.00
		600	65	

S. No.	Manufacturer	Rating (kW)	Details as on 31st March 2006	
31	Wind Master	200	1	0.20
32	Windmatic	55	30	1.65
33	Wind Power	330	29	9.57
	TOTAL		**10825**	**5340.96**

Source: [24]

At present, efforts are being made to develop a low cost, indigenous, horizontal axis Wind Energy Generator (WEG) of 500 kW rating. The WEG will have a two bladed rotor and the tower will be a tubular tower with guys. The organizations contributing in the development of the WEG are (i) National Aerospace Laboratory (NAL), (ii) Structural Engineering Research Centre (SERC), (iii) Sangeet Group of Companies, and (iv) Center for Wind Energy Technology (C-WET). It will be specially suited for Indian wind conditions i.e. relatively low wind speeds and dusty environment. It is further learnt that this WEG may cost almost 50% as compared to the other WEGs of the same rating commercially available in India. The WEG is nearing completion and likely to be completed by April-2007 [24].

Wind in India is dominated by the strong south-west summer monsoon, which starts in May–June, when cool, humid air moves towards the land and the weaker north-east winter monsoon, which starts in October, when cool, dry air moves towards the ocean. During the period March to August, the wind is uniformly strong over the whole Indian Peninsula, except the eastern peninsular coast. Wind speeds during the period November to March are relatively weak, though higher winds are available during a part of the period on the Tamil Nadu coastline.

In order to tap the potential of wind energy sources, there is a need to assess the availability of the resources spatially. A Wind Resource Assessment Programme was taken up in India in 1985 [19]. Around 1150 wind monitoring/mapping stations were set up in 25 states and Union Territories (UTs) for this purpose. Over 200 wind monitoring stations in 13 states and UTs with annual mean wind power density greater than 200 W/m^2 at a height of 50 m above the ground level show wind speeds suitable for wind power generation [25]. The wind power density at a height of 50 m above the ground level is depicted in Figure 3.

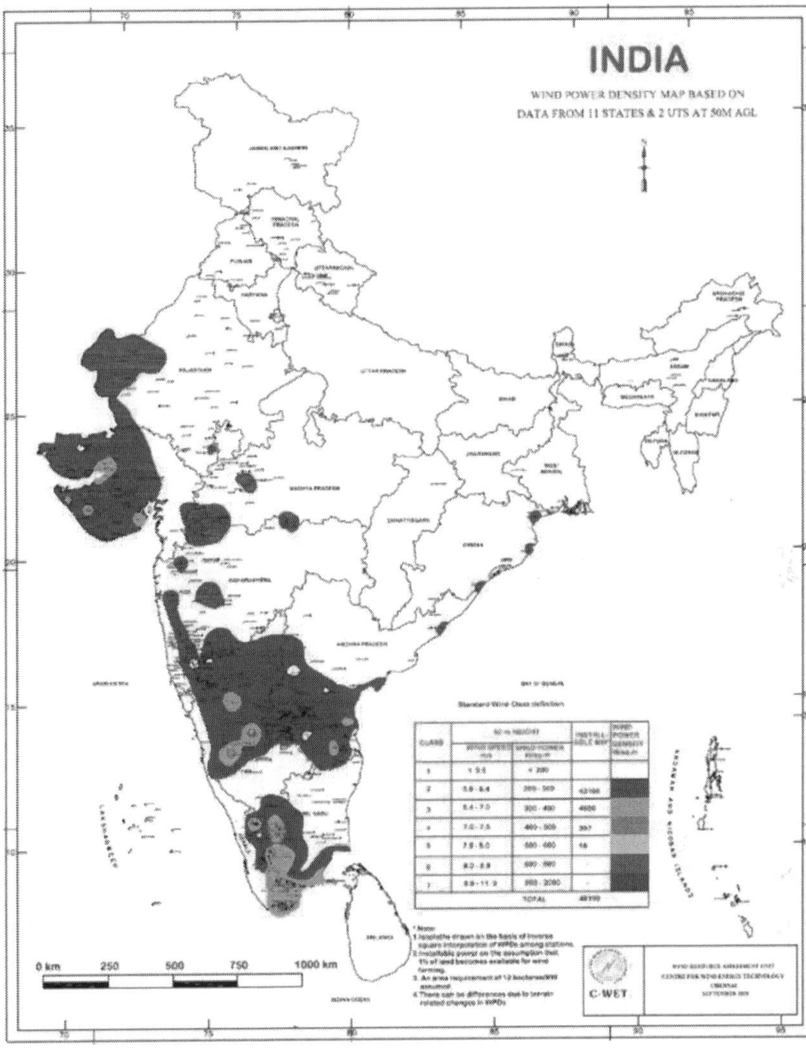

Figure 3 Wind power potential in India (Source: Centre for Wind Energy Technology (C-WET), Government of India).

On a regional basis, more detailed assessments have been done. Ramachandra and Shruthi [26] employed a geographical information system (GIS) to map the wind energy resources of Karnataka state and analyzed their variability considering spatial and seasonal aspects. A spatial data base with data of wind velocities has been developed and used for evaluation of the theoretical potential through continuous monitoring and mapping of the wind resources. The study shows that the average wind velocity in Karnataka varies from 0.9 m/s in Bagalkote to 8.3 m/s in Chikkodi during the monsoon season. Agroclimatic zone wise analysis shows that the northern dry zone and the central dry zone are ideally suited for harvesting wind energy for regional economic development.

Onshore wind power potential in the country has been assessed at 45 GW assuming 1% of land availability for wind power generation in the potential areas [27]. However, it is estimated that a penetration (supply fraction) of wind power on a large grid can be as much as 15–20% without affecting grid stability due to requirement of reactive power [28]. Therefore, at present, it is not technically feasible to exploit the full wind power potential in view of total installed power-generating capacities from conventional power generating methods including hydro-electric power plants in different states. Considering a maximum of 20% penetration of existing capacities of the grids through wind power in the potential states, technical potential for grid interactive wind power is presently limited to only 13 GW [25]. Total technical potential for wind power in the country is expected to increase with augmentation of grid capacity in potential states. Table 2presents a state wise break-up of the estimated technical potential along with wind power installed capacity as on 30 September 2006. One should note that Tamil Nadu has already surpassed the presumed technical potential, indicating that it may be underestimated for India as a whole.

Table 2 State wise gross wind power potential, technical potential and cumulative installed capacity in India up to 30.09.2006

State	Gross potential (MW)	Technical potential (MW)	Installed capacity (MW)
Andhra Pradesh	8275	1750	121
Gujarat	9675	1780	376
Karnataka	6620	1120	688
Kerala	875	605	2
Madhya Pradesh	5500	825	53
Maharashtra	3650	3020	1242
Orissa	1700	680	2

State	Gross potential (MW)	Technical potential (MW)	Installed capacity (MW)
Rajasthan	5400	895	386
Tamil Nadu	3050	1750	3148
West Bengal	450	450	2
Total (All India)	**45195**	**12875**	**6018**

Source: [25]

The original impetus to develop wind energy in India came in the early 1980s from the government, when the Commission for Additional Sources of Energy had been set up in 1981 and upgraded to the Department of Non-Conventional Energy Sources in 1982. The setup of these institutions was due to the wish to encourage a diversification of fuel sources away from the growing demand for coal, oil and gas required to meet the demand of the country's rapid economic growth [29]. A market-oriented strategy was adopted from inception, which has led to the successful commercial development of the technology. The broad based national programme included wind resource assessment; research and development support; implementation of demonstration projects to create awareness and opening up of new sites; involvement of utilities and industry; development of infrastructure capability and capacity for manufacture, installation, operation and maintenance of wind power plants; and policy support.

The Ministry of Non-Conventional Energy Sources (MNES) which was set up in 1992 has been providing support for research and development, survey and assessment of wind resources, demonstration of wind energy technologies and has also taken fiscal and promotional measures for implementation of private sector projects [30, 31]. India now has a fairly well-developed and growing wind power industry with a number of Indian companies involved in manufacturing of wind turbines. These companies have tied up with foreign wind power industries for joint venture/licensed production in India, for their market shares see Table 1. Wind turbines up to 2 MW are presently manufactured in India [25]. Figure 4presents the cumulative capacity of wind power installed in India over time.

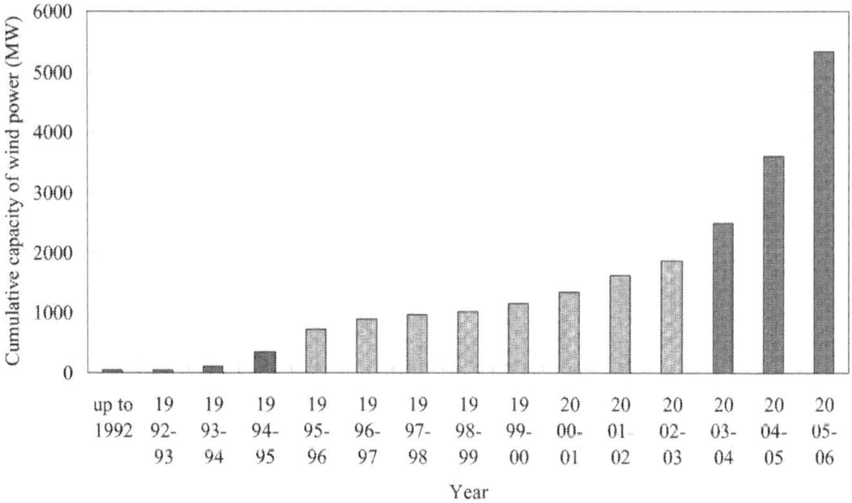

Figure 4 Development of wind power capacity in India over time (Source: MNES Annual Reports).

A notable feature of the Indian programme has been the interest among private investors/developers in setting up of commercial wind power projects. This is due to a range of fiscal incentives provided by the Indian government such as 80% accelerated depreciation, tax holiday for power generation projects, concessional customs and excise duty as well as liberalized foreign investment procedures [25, 29, 31]. The Indian Renewable Energy Development Agency (IREDA) provides concessional loans. Current interest rates are 9.5% for a maximum repayment period of 10 years and 9.0% for a maximum repayment period of 8 years [25]. Table 3 presents the summary of key central government incentives for wind power projects in India.

Table 3 Financial and fiscal incentives for wind power projects in India

Type of incentive	Description	Rate
I. Indirect Taxes	i) Wind operated electricity generators upto 30 kW and wind operated battery chargers upto 30 kW.	5%
	ii) Parts of wind operated electricity generators for manufacturer of wind operated electricity generators, namely: Special bearing, Gear Box, Yaw components,	5%

Type of incentive	Description	Rate
	Wind turbine controllers.	
	Sensors, Brake hydraulics, Flexible coupling, Brake callipers.	25%
	iii) Blades for rotor of wind operated electricity generators for the manufacturers or the manufacturers of wind operated electricity generators.	5%
	iv) Parts for the manufacturer or the maintenance of blades for rotor of wind operated electricity generation.	5%
	v) Raw materials for manufacturer of blades for rotor of wind operated electricity generators.	5%
II. Excise Duty	Devices/Systems exempted from Excise Duty:	
	i) Wind operated electricity generator, its components and parts thereof including rotor and wind turbine controller.	
	ii) Water pumping wind mills, wind aero-generators and battery chargers. [Notification No.6/2002 dated 01/03/2002 (S.No.237 non-conventional devices/systems)]	
III. Sales Tax	Exemption/reduction in Central Sales Tax and General Sales Tax are available on sale of renewable energy equipment in various states.	
	i) 80% Accelerated Depreciation on specified Non-conventional Renewable Energy devices/systems (including wind power equipment) in the first year of installation of the projects.	
	ii) Tax Holiday on Power Projects.	

Sources: Ministry of Non-conventional Energy Sources, Government of India, New Delhi

The MNES has issued guidelines to all state governments to create an attractive environment for the export, purchase, wheeling and banking of electricity generated by wind power projects. The guidelines include the promotion of renewables including wind energy through preferential tariffs and a minimum obligation on distribution companies to source a certain share of electricity from renewable energy. However, only a subset of states is actually complying with these guidelines. The State Electricity Regulatory Commissions (SERCs)of Andhra Pradesh, Madhya Pradesh, Karnataka and Maharashtra provide preferential tariffs for wind power. Maharashtra, Andhra Pradesh, Karnataka, Madhya Pradesh and Orissahave enacted the renewables obligation on distributors. The problem with incentives on the state level is that they vary erratically and thus cannot be taken for granted by project developers (see Table 4 for the case of Rajasthan).

Table 4 Policy of the state of Rajasthan for sale of power from wind installations

Period	**Policy of the state of Rajasthan for sale of power from wind installations**
March 1999 – February 2000	Electricity could be purchased at INR 2.75 (US$ 0.061/kWh) with just 2% wheeling charges along with sales tax incentives. The developer was allowed to bank electricity for one year.
February 2000 – April 2003	Electricity could be purchased at INR 3.03 (US$ 0.067/kWh) while the wheeling charges were kept same at 2%. The provision for banking for 12 months was limited to end of financial year only (March 31). If the banking period is exhausted and the electricity was not sold out by then, the state power utility would buy balance amount of electricity at 60% of the agreed purchase price.
April 2003 – October 2004	Electricity could be purchased at INR 3.32 (US$ 0.073/kWh). The wheeling charges were drastically increased from 2% to 10% for the volume of electricity supplied to the grid. The banking period was reduced from 12 months to the end of calendar year (December 31).
October 2004 – Onwards	The purchase price was reduced from INR 3.32/kWh (US$ 0.073/kWh) to INR 2.91/kWh (US$ 0.064/kWh) which is 13% lower than the previous power policy.

Source: [33]

The main attraction for private investment is the fact that owning a wind turbine assures a profitable power supply compared to the industrial power tariff, which is kept artificially high to cross-subsidize electricity tariffs for farmers. Therefore, clusters of individually owned wind turbines appear to substitute grid electricity. More than 97% of investment in the Indian wind sector is provided from the private sector [25]. However, the impending liberalization under the Electricity Act 2003 may take away this key incentive if industrial power users can procure electricity at competitive rates.

The Clean Development Mechanism under the Kyoto Protocol allows developing countries to generate emission credits (CERs) for industrialized countries by GHG emission reduction projects such as wind power. The sale of CERs could help to accelerate wind power development in India. We assess the theoretical CDM potential of wind power projects in India before discussing whether at the current market situation such projects could become attractive.

Results: CDM Potential of Wind Power Projects in India

Considerable variation has been observed in the reported values of the PLF of the wind power plants in the CDM Project Design Documents (Table 5). Therefore, in this study to estimate the CDM potential of wind power projects in India the PLF of the wind power plants have been taken as 25%. There are five regional grids within the country – the Northern, Western, Southern, Eastern and North-Eastern. Therefore, the CO_2 emissions mitigation potential through wind power projects in India is estimated on the basis of the regional grids, whose emission factors have been calculated by the Central Electricity Authority (CEA) of the Government of India in 2006. Table 6 presents the estimated values of CDM potential through wind power projects in India on the basis of the regional baselines.

Table 5 Additionality Test of Indian projects on wind power

Title	Methodology	Investment Analysis	Barrier Analysis	Investment and Barrier Analysis	Identification of alternatives	Institutional /Regulatory Barriers	Technology Barriers	Common Practice Analysis	Impact of CDM registration	Remarks
Nagda Hills (6.25 MW) Wind Energy Project	AMS-I.D.	√	√, I/IR/PP	√	√	√	×	√	√	PLF = 29%; IRR without CDM ~ 9.8%; IRR with CDM ~ 13.5%.
12.3 MW wind energy project in Tamil nadu, India	AMS-I.D.	√	√, I/IR/T/PP	√	×	√	√	√	√	PLF = 22%; IRR without CDM ~ 12.9%; IRR with CDM ~ 13.4%.
14.8 MW small-scale grid connected wind power project in Jaisalmer state Rajasthan	AMS-I.D.	√	√, I/PP	√	×	×	×	√	×	The PLF was considered as 25% before the WEGs started operating, it was later found out to be less than 18%. IRR of the project activity reduced to less than 10% after the execution of the project.
Bundled Wind power project in Jaisalmer, Rajasthan (58.2 MW)	ACM2	√	√, I/IR/PP	√	√	×	×	√	√	PLF at 22.28% (IRR = 9.2% without CDM and 14.6% with CDM); PLF at 25.28% (IRR = 11.0% without CDM and IRR = 17.1% with CDM).
Bundled wind power project in Chitradurga (Karnataka in India) managed by Enercon (India) Ltd. (16.8 MW)	ACM2	√	√, I/PP	√	√	×	×	√	√	PLF at 26% (IRR = 9.5% without CDM and 11.5% with CDM); PLF at 30% (IRR = 14.8% without CDM and IRR = 17.4% with CDM).
3.75 MW Small Scale Grid Connected "Demonstration Wind Farm Project" at Chalkewadi, District Satara, State Maharashtra	AMS-I.D.	×	√, I/IR/T	×	×	√	√	×	×	PLF = 18 – 20%; The investor saw CDM revenue as a risk mitigation against these uncertainties.
11.35 MW Grid Connected Wind Electricity Project at Pohra (Rajasthan)	AMS-I.D.	×	√, I/IR/T	×	×	√	√	×	×	PLF = 20 – 22%; The investor saw CDM revenue as a risk mitigation against these uncertainties.
10.6 MW wind farm at Village Badabagh, District Jaisalmer, Rajasthan.	AMS-I.D.	×	√, I/IR/T	×	×	√	√	×	×	PLF varies from 14.7 to 22.5%.
56.25 MW bundled wind energy project in Tirunelveli and Coimbatore districts in Tamilnadu	ACM2	√	√, I/PP	√	√	×	×	√	√	PLF = 14 – 17.5%; IRR = 10.1% without CDM and IRR = 12.1% with CDM.
5 MW Wind Project at Baramsar and Soda Mada, Jaisalmer, Rajasthan	AMS-I.D.	×	√, I/IR/T	×	×	√	√	×	×	Investment barriers exists.
7.5 MW wind farm of REI Agro Ltd. at Soda-Mada in the state of Rajasthan	AMS-I.D.	×	√, I/IR	×	×	√	×	×	×	Investment barriers exists.
11.25 MW wind power project in Dhule, Maharashtra, India	AMS-I.D.	√	√, I	√	×	×	×	×	×	IRR without CDM ~ 14.17%; IRR with CDM ~ 21.59% which is above the acceptable bench mark IRR of 15.06%.

Wind Electricity Generation at Erakandurai, Dist:Tirunavalli by M/s GHCL Ltd	AMS-I.D.	√	√, I	√	×	×	×	×	×	PLF at 22.83%; IRR = 11.54% without CDM and 14.70% with CDM. Similarly, with PLF at 21.43%; IRR = 9.72% without CDM and 12.58% with CDM.
125 MW wind power project in Karnataka	ACM2	√	√, I/PP	√	√	√	√	√	√	IRR = 7.36% without CDM revenues and 7.87% with CDM revenues.
Generation of electricity from 6.25 MW capacity wind mills by Sun-n-Sand Hotels Pvt. Ltd at Soda Mada Rajasthan	AMS-I.D.	√	√, I/IR/T/PP	√	√	√	√	√	√	PLF = 17 – 19%; IRR without CDM ~ 12.45%; IRR with CDM ~ 14.81%
Generation of electricity from 4 MW capacity wind mills by Sun-n-Sand Hotel group at Supa, Maharashtra	AMS-I.D.	√	√, I/IR/T/PP	√	√	√	√	√	√	PLF = 20%; IRR without CDM ~ 13.76%; IRR with CDM ~ 16.53%
Generation of electricity from 2.5 MW capacity wind mills by Gujarat JHM Hotels Ltd. Ltd at Soda Mada, Rajasthan	AMS-I.D.	√	√, I/IR/T/PP	√	√	√	√	√	√	PLF = 17 – 19%; IRR without CDM ~ 10.57%; IRR with CDM ~ 12.93%
Generation of electricity from 1.2 MW capacity wind mills by Sun-n-Sand Hotels Pvt. Ltd at Satara, Maharashtra	AMS-I.D.	√	√, I/IR/T/PP	√	√	√	√	√	√	PLF = 22 – 25%; IRR without CDM ~ 16.84%; IRR with CDM ~ 19.86%.
15.4 MW wind farm at Satara District, Maharashtra*	ACM2	×	√, I/IR	×	√	√	×	×	√	PLF = 19.24%
4.2 MW Wind power project in Maharashtra, by Bharat Forge Limited*	AMS-I.D.	√	√, I/IR/T	√	×	√	√	×	√	PLF = 13.09 – 23.96%; IRR without CDM ~ 14.3%; IRR with CDM ~ 16.4%.

I: Investment barrier; T: Technological barrier; I/R.: Institutional and/or regulatory barriers; PP: Barriers due to the prevailing practice. *Reg. request
Source: [33]

Table 6 Annual gross and technical CO_2 emissions mitigation potential through wind power projects in India

State	**Region**	**Baseline* (kg CO_2/kWh)**	**Annual electricity generation (TWh)**		**Annual CO_2 emissions mitigation potential (million tonnes)**	
			Gross	Technical	Gross	Technical
Andhra Pradesh	Southern	0.86	18.1	3.8	15.6	3.3
Gujarat	Western	0.89	21.2	3.9	19.0	3.5
Karnataka	Southern	0.86	14.5	2.5	12.5	2.1
Kerala	Southern	0.86	1.9	1.3	1.6	1.1
Madhya Pradesh	Western	0.89	12.0	1.8	10.8	1.6
Maharashtra	Western	0.89	8.0	6.6	7.2	5.9
Orissa	Eastern	1.04	3.7	1.5	3.9	1.5
Rajasthan	Northern	0.75	11.8	2.0	8.9	1.5
Tamilnadu	Southern	0.86	6.7	3.8	5.8	3.3
West Bengal	Eastern	1.04	1.0	1.0	1.0	1.0
All India			99.0	28.2	86.2	24.9

*Source: [35]

We now do a sensitivity analysis with regards to additionality determination. The case of lax additionality assumes that all wind power projects submitted are registered. The median case assumes that the rejection rate remains at the current level (2 out of 18 projects, i.e. 11%). The case of stringent additionality assumes that 50% of the projects are registered. In the lax additionality case, gross annual CER potential of wind power in India reaches 86 million. Similarly, based on the technical potential of wind power projects in India the CDM potential has been estimated as 25 million tonne. Among all the states in India, Gujarat has the largest CO_2 emissions mitigation potential through wind power (19 million tonne) followed by Andhra Pradesh (15.6 million tonne), Madhya Pradesh (10.8 million tonne), Karnataka (12.5 million tonne), Rajasthan (8.9 million tonne), and so on (Table 6). The annual electricity generation by wind power projects based on the gross and technical potential is also given in Table 6. With 25% PLF of wind power projects the annual gross electricity generation potential has been estimated at 99 TWh whereas the annual technical electricity generation potential has been estimated at 28 TWh.

Table 7 presents the projected values of the cumulative capacity of wind power and likely CER generation using the logistic model described in the Methods section while Figure 5 shows the development over time. It may be noted that with the current trend of dissemination of wind power projects in the country, around 22 GW capacity could be installed up to the end of first crediting period in the SS scenario whereas in the OS scenario 36 GW capacity could be installed. Up to the the year 2020, more than 44 GW capacity of the wind power projects are expected to be installed that would generate 87 million CERs.

Table 7 Projected values of the cumulative capacity of wind power and associated CER generation

Year	Projected values of the cumulative capacity (GW)		Projected values of the annual electricity generation (TWh)		Projected values of the annual CER generation (million CERs)	
	SS	OS	SS	OS	SS	OS
2008	10	23	21	50	18	43
2012	22	36	48	78	41	67
2016	35	42	76	92	65	79
2020	41	44	91	97	78	83

*Baseline 860 g CO_2/kWh

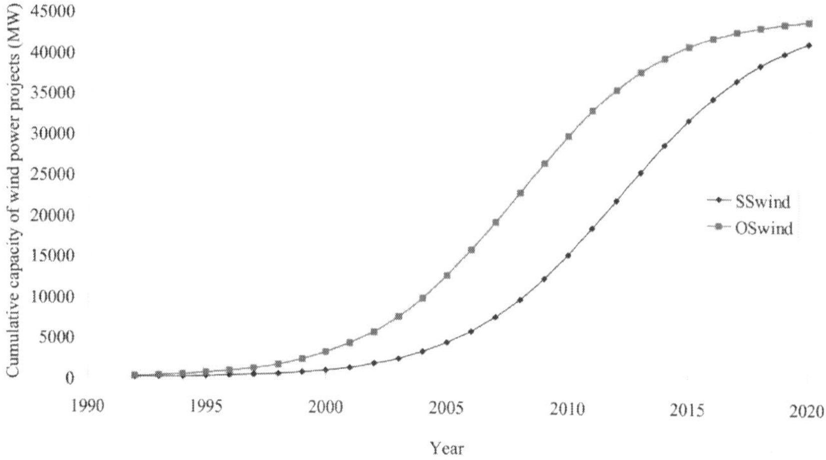

Figure 5 Realistic CDM potential for wind power until 2020.

Discussion: How the CDM could be applied to the Diffusion of Wind Power Projects?

The CDM was slow to take off as after the Marrakech Accords of 2001 it took another three years to define the bulk of the rules. The CDM Executive Board (EB) which is the body defining the CDM rules surprised many observers by taking a rigorous stance on critical issues such as baseline and additionality determination (see below). Once the key rules were in place, a "gold rush" happened in 2005 and 2006. Over 1500 projects were submitted with an estimated CER volume of about 1.5 billion. However, the volume share of renewable energy projects has been less than expected due to the high attractiveness of projects reducing industrial gases and methane from waste. Out of the 1478 CDM projects submitted to the EB, 456 projects had been registered by the EB till 20th December, 2006 [32, 33]. 183 CDM projects related to wind energy of which 47 have been registered, 9 requested registration and 127 were at the validation stage [33]. Figure 6presents the status of the wind power projects from India. Out of the 89 projects submitted to the UNFCCC, 18 projects had been registered and two projects had submitted the request for registration. 67 projects were at the validation stage whereas 2 projects had been rejected by the EB.

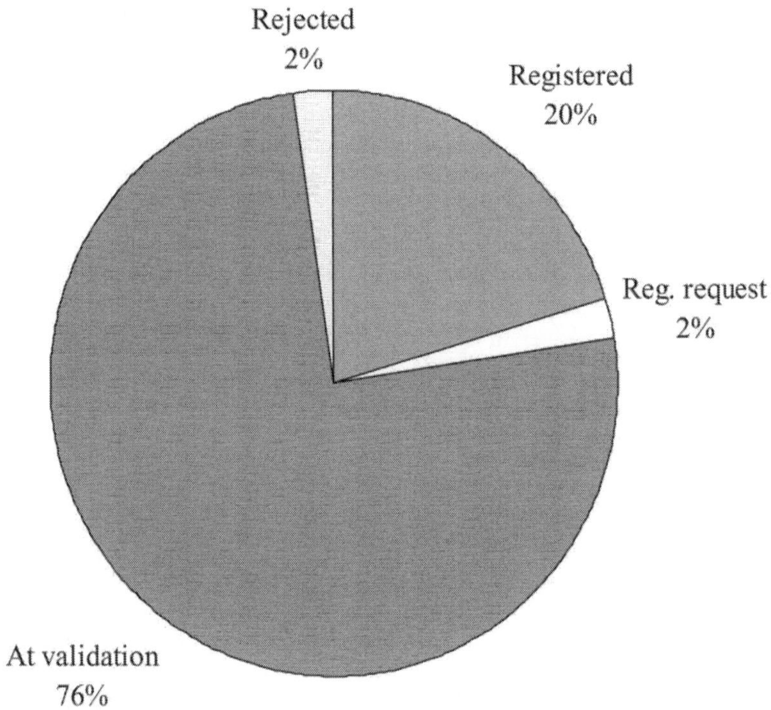

Figure 6 Status of the wind power projects from India till 20th December 2006 (Source: [33]).

Baseline

The quantification of GHG benefits of a CDM project is done by means of a "baseline". A baseline describes the (theoretical) emissions that would have occurred in case the CDM project was not implemented. The amounts of CERs that can be earned by the project are then calculated as the difference of baseline emissions and project emissions. The CO_2 emissions mitigation benefits associated with a wind power project depend upon the amount of electricity saved. To estimate the CDM potential of wind power project in the country, the approved consolidated baseline methodology for grid-connected electricity generation from renewable sources ACM0002 (Version 06) has been used. For the small scale CDM (SSC) projects, the small scale methodology AMS-I.D. "Grid connected renewable electricity generation" in its version of 23rd December 2006 [34] can be used which explicitly mentions wind power for electricity generation. In India, most of the wind power projects are grid connected and substitute grid electricity. Therefore, for such systems, the baseline is the kWh produced by the renewable generating unit multiplied by an emission coefficient (measured in g CO_2eq./kWh) calculated in a transparent and conservative manner. This coefficient is 800 g CO_2eq./kWh for a grid where all generators use exclusively fuel oil and/or diesel fuel, whereas it is the weighted average of the so-called operating margin (emission factor of all thermal power plants serving the grid) and build margin (emission factor of the most recently built plants that provide 20% of the grid's electricity). For wind power, the weight of the operating margin is 0.75 while the build margin is weighted at 0.25. Alternatively, project developers can use the weighted average emissions of the current generation mix but this will always be less than the emission factor derived previously and thus unattractive. For intermittent and non-dispatchable generation types such as wind and solar photovoltaic, ACM0002 allows to weigh the operating margin (OM) and build margin (BM) at 75% and 25%, respectively, however, in this study we have used combines margin by using equal weights for OM and BM as given in CEA document [35].

Additionality

To maintain the environmental integrity of the Kyoto Protocol, CERs are given only for "additional" activities that would otherwise not be expected to occur [36]. Therefore, any CDM project requires careful analysis of additionality. This has probably been the most contentious point in the development of the CDM and also resulted in great confusion amongst project developers [37, 38]. The Kyoto Protocol stops short of requiring project proponents to show strict financial additionality – that the CDM revenue makes an uneconomic project economic – and left scope for the CDM EB to refine the demonstration of additionality. The EB subsequently took a fairly strict interpretation of additionality and developed an additionality tool which formally is voluntary but which has become de facto mandatory as it was incorporated in most baseline methodologies. The additionality tool requires an investment analysis and/or a barrier analysis to determine whether the CDM project is the most attractive realistic alternative. This means that the project can be profitable and additional as long as developers can show that another project type was even more profitable.

It is estimated that wind power in many countries is already competitive with fossil fuel and nuclear power if social/environmental costs are considered [28]. However, in India, in terms of costs per kWh in grid-connected areas, costs of wind power are higher than electricity provided by a coal plant projects thus be additional at any rate. The unit cost of electricity generation is 0.05 €/kWh for coal and 0.06 €/kWh for fuel oil based system whereas in case of wind, the unit cost of electricity generation is 0.07 €/kWh in the best locations. The problem with this reasoning is that if wind projects are used to displace expensive grid electricity for industrial consumers (priced at 0.09 €/kWh [39]), they are invariably the most attractive alternative unless they are built in locations with low wind speed. The situation for wind projects that supply to the grid at the state-guaranteed feed-in tariff is less clear; the attractiveness depends on the level of the tariff.

As the investment test will not be passed by most wind projects (or only if they omit the tax incentives, as has been done by a project that achieved registration), project developers will use the barrier test. The barrier of higher capital cost compared to fossil fuel power plants is not really credible due to the abundance of capital for wind power in India and thus is mentioned only rarely. More credible barriers are low capacity utilization factor, and possible reduction in feed-in tariffs. The former depends on the siting of the project. The latter is very important as shown by the policy of Rajasthan (see Table 3) and other states. In 2001, Tamil Nadu Electricity Board (TNEB) changed its policy and froze the power purchase tariff for wind energy at Rs 2.70 per kWh with no escalation till 2006 and had informed that this power purchase tariff would be reviewed at 2006 and a new tariff would be fixed then. This was a major barrier for establishing new wind farms as other renewable energy plants continued to get a higher tariff. For instance, the power purchase tariff for electricity from an industrial waste/municipal waste based generation was Rs 3.49 for the year 2005 as against Rs 2.70 for wind energy. This policy encourages investors to invest in other renewable energy plants. Reduction in power purchase tariff was a major investment barrier. Moreover, feed-in-tariffs may be replaced by the Availability Based Tariff (ABT) in which the generators with firm delivery of power against commitment will start getting more prices for the generated power, whereas wind power producers cannot guarantee supply of electricity and will be thus receive lower rates. For the projects that substitute grid electricity at industrial tariffs, there is the risk that the wind power benefit will melt down as liberalization permits industrial electricity consumers to choose the supplier in a competitive environment. Some projects have also highlighted the technological risks associated with new types of wind turbines. Lack of familiarity and experience with such new technologies can lead to perceptions of greater technical risk than for conventional energy sources.

Doing the investment test – case study

A 125 MW wind project in Karnataka calculated an IRR of 7.3%. At that rate, the project would clearly be unattractive for an investor. However, the picture changes if one analyzes the project more closely. If one uses industry averages for the investment cost (Rs 5 crore per MW), the IRR is 11%. If one includes the accelerated depreciation of 80% in the first year and the 10 year income tax holiday, the IRR reaches 22% (personal communication by Mr. Sanjeev Chadha). It would be difficult to find serious alternatives that are more attractive. Nevertheless, the project was registered by the EB.

Table 5 presents the additionality arguments of Indian wind power projects. 14 projects out of 20 have carried out investment and barrier analysis for the justification of additionality whereas 6 projects carried out the barrier analysis only. An assessment of the PDD's indicates that the investment analysis is not convincing in most of the cases. Two wind projects from India were rejected due to lack of additionality. The rejection was mainly due to the following statement in the annual report of the company that had invested in the projects: "The project is extremely beneficial on a standalone basis and has a payback period of three years with an internal rate of return in excess of 28 per cent. In addition to hedging Bajaj Auto's power costs, this investment also provides sales tax incentives and an income tax shield" [40].

Monitoring

For wind power plants, monitoring is easy – you just meter the electricity generated and sold to the grid.

CONCLUSION

Our estimates indicate that, there is a vast theoretical potential of CO_2 mitigation by the use of wind energy in India. On the basis of available literature, the gross potential of wind power is more than 45,000 MW. The annual CER potential of wind power in India could theoretically reach 86 million tonnes. Under more realistic assumptions about diffusion of wind power projects based on past experiences with the government-run programmes, annual CER volumes by 2012 could reach 41 to 67 million and 78 to 83 million by 2020. The projections based on the past diffusion trend indicate that in India, even with highly favorable assumptions, the dissemination of wind power projects is not likely to reach its maximum estimated potential in another 15 years. CDM could help to achieve the maximum utilization potential more rapidly as compared to the current diffusion trend if supportive policies are introduced.

METHODS

CO_2 emissions mitigation potential of a windmill

The power output of a windmill essentially depends on the site/location specific parameters (such as wind speed, air density, etc.) and design parameters (such as coefficient of performance of the wind rotor, swept area of the rotor, cut-in, cut-out and rated wind speed of the rotor, etc.) of the windmill. Therefore, the annual useful energy, AUE_{wind}, delivered by a windmill can be estimated as [20]

$$AUE_{wind} = 8760\gamma\int_{v_{ci}}^{v_{co}} P(v)F(v)dv \tag{1}$$

where γ represents the windmill turbine mechanical availability factor accounting for downtime during maintenance etc., P(v) the power produced by the windmill at wind speed v (in m/s), F(v) the Weibull probability distribution function, v_{ci} the cut-in wind speed and v_{co} the cut-out wind speed of the windmill.

The power produced by the windmill at wind speed v may be expressed as [41]

$$P(v) = \frac{1}{2}C_p\rho_a Av^3 \tag{2}$$

where C_p represents the coefficient of performance of the wind rotor, ρ_a the density of air, A the swept area of the rotor and v the wind speed.

The variation in wind speed at a location is often described by the Weibull probability distribution function F(v). The Weibull probability density function is given by the following expression [20, 42]

$$F(v) = \left(\frac{k}{c}\right)\left(\frac{v}{c}\right)^{(k-1)} e^{-\left(\frac{v}{c}\right)^k} \tag{3}$$

where k represents the shape parameter and c the scale parameter.

Substituting the values of P(v) and F(v) from Eqs. (2) and (3) into Eq. (1) the annual useful energy (in kWh) delivered by the windmill can be expressed as [43]

$$AUE_{wind} = 4.38\left(\frac{\gamma C_p \rho_a A k}{c^k}\right)\int_{v_{ci}}^{v_{co}} v^{(k+2)} e^{-\left(\frac{v}{c}\right)^k} dv$$

(4)

The annual gross CO_2 emissions mitigation potential of a windmill essentially depends upon the annual electricity saved by the windmill and the CO_2 emission factor of the electricity.

$$GCE_{wind} = 4.38\left(\frac{\gamma C_p \rho_a A k}{c^k (1-l)}\right)\left[\int_{v_{ci}}^{v_{co}} v^{(k+2)} e^{-\left(\frac{v}{c}\right)^k} dv\right] CEF_e$$

(5)

where l (in fraction) represents the electrical transmission and distribution losses of the grid and CEF_e the baseline CO_2emission factor.

For a given capacity of a wind power project the CO_2 emissions mitigation potential can be estimated as:

$$GCE_{wind} = (8760\ PLF_{wind}\ P_{wind})\ CEF_e$$

(6)

where P_{wind} (in MW) represents the capacity of wind power project, PLF_{wind} (in fraction) the plant load factor of the wind power project, CEF_e the CO_2 emission factor of electricity. The term inside the second bracket of the right hand side of Eq. (6) is the annual amount of electricity saved by the wind power project.

Estimation of Diffusion of Wind Power Projects in India

The diffusion of a technology measured in terms of the cumulative number of adopters usually conforms to an exponential curve [44] as long as the new technologies manage to become competitive with incumbent technologies. Otherwise, the steep section of the curve would never be reached because technology use falls back to zero at the removal of subsidies [45]. The exponential growth pattern may be of three types – (i) simple exponential, (ii) modified exponential, and (iii) S-curve. Out of these three growth patterns, the simple exponential pattern is not applicable for the dissemination of renewable energy technologies, as it would imply infinite growth [46]. The modified exponential pattern (with a finite upper limit) is more reasonable but such a curve may not match the growth pattern in the initial stage of diffusion [47, 48]. Empirical studies have shown that in a variety of situations the growth of a technology over time may conform to an S-shaped curve, which is a combination of simple and modified exponential curves [49, 50]. The S-shaped curves are characterized by a slow initial growth, followed by rapid growth after a certain take-off point and then again a slow growth towards a finite upper limit to the dissemination [51]. However, a logistic model is used to estimate the theoretical cumulative capacity of wind power projects at different time periods.

As per the logistic model, the cumulative capacity, P(t), of the wind power projects disseminated up to a particular period (t^{th} year) can be expressed as [49]

$$P(t) = P_{max}\left[\frac{e^{(a+bt)}}{1+e^{(a+bt)}}\right] \tag{7}$$

where P_{max} represents the estimated maximum utilization potential of the renewable energy technology in the country. The regression coefficients a and b are estimated by a linear regression of the log-log form of equation as given below.

$$ln\left[\left(\frac{P(t)}{P_{max}}\right) \Big/ \left\{1-\left(\frac{P(t)}{P_{max}}\right)\right\}\right] = a + bt \tag{8}$$

Figure 7 represents the projected time variation of the cumulative capacity of wind power using the logistic model considered in the study. Two cases such as business as usual or standard scenario (SS) and optimistic scenario (OS) are presented. The values of the regression coefficients using a logistic model have been estimated by regression of the time series data for the installation of wind power (Figure 4) extracted from the annual reports of the MNES [25]. In the optimistic scenario it is assumed that, in the past, if the diffusion of wind power would have been driven by the market forces instead of subsidies then the cumulative capacity of installation of wind power would be three times more than the actual level [52, 53]. Our results indicate that in India, even with highly favourable assumptions, the dissemination of wind power projects is not likely to reach its maximum estimated potential in another 15 years. But all these time periods are not relevant for the CDM whose current endpoint is 2012 and which may only be able to live longer if post-2012 negotiations retain an emission target based policy regime. However, CDM could be used as a tool to foster the dissemination of wind power projects in the country. It could accelerate the diffusion process.

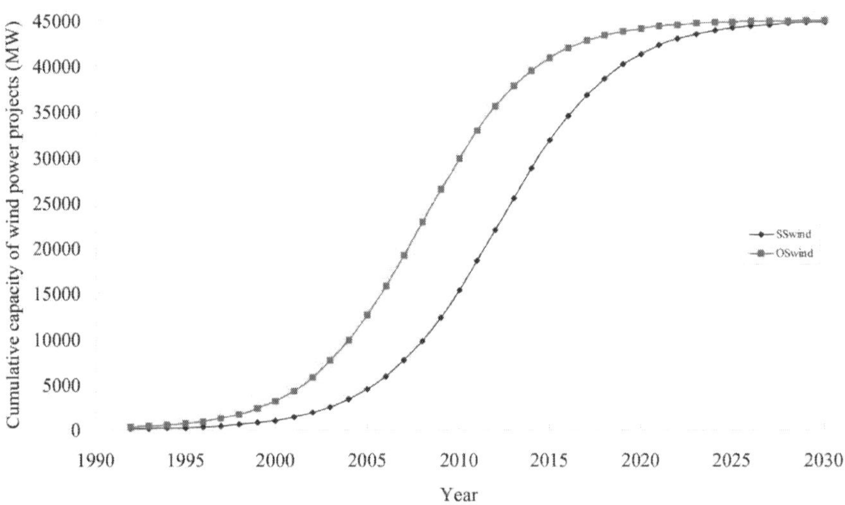

Figure 7 Time variation of cumulative capacity of wind power in India using logistic model.

ACKNOWLEDGEMENTS

The financial assistance provided by the e8 Network Secretariat, Canada to the first author (Pallav Purohit) in the form of e8 Postdoctoral Scholarship on Sustainable Energy Development is gratefully acknowledged.

COMPETING INTERESTS

Axel Michaelowa is working for CDM consultancy Perspectives, which has not been involved in wind power CDM projects in India

AUTHORS' CONTRIBUTIONS

Both authors have worked on all parts of the article, with Pallav Purohit focusing on the modeling and AM focusing on CDM issues.

REFERENCES

1 IEA: *World Energy Outlook 2006.* International Energy Agency (IEA), Paris; 2006.

2 Planning Commission: *Tenth Five-year Plan (2002–07).* Govt of India, New Delhi; 2002.

3 Ghosh S: **Future demand of petroleum products in India.** *Energy Policy* 2006, **34:** 2032. 10.1016/j.enpol.2005.02.007

4 Ministry of Power (MOP), Govt. of India, New Delhi; 2006. [http://powermin.nic.in]

5 People's Daily: **China's power generating capacity tops 622 million kilowatts.**[http://english.peopledaily.com.cn/200701/23/eng20070123_343745.html] January 23, 2007.

6 **GWEC position paper on the IEA World Energy Outlook 2006**Global Wind Energy Council (GWEC), Brussels; 2006. [http://www.gwec.net/fileadmin/documents/Publications/GWEC_Position_Paper_on_IEA_WEO_2006.pdf]

7 Jagadeesh A: **Integration of wind farms into the public power system in India.** *Journal of Wind Engineering and Industrial Aerodynamics* 1988, **27:** 433. 10.1016/0167-6105(88)90056-6

8 **Global wind energy markets continue to boom – 2006 another record year**Global Wind Energy Council (GWEC), Brussels; 2007. [http://www.gwec.net/uploads/media/0702_PR_Global_Statistics_2006.pdf]

9 **Global Wind Power Continues Expansion**Global Wind Energy Council (GWEC), Brussels; 2005. [http://www.gwec.net]

10 Wüstenhagen R, Markard J, Truffer B: **Diffusion of green power products in Switzerland.** *Energy Policy* 2003, **31:** 621. 10.1016/ S0301-4215(02)00147-7

11 Jacobsson S, Lauber V: **The politics and policy of energy system transformation – explaining the German diffusion of renewable energy technology.** *Energy Policy* 2006, **34:** 256. 10.1016/ j.enpol. 2004.08.029

12 Scholz S: **The POETICs of industrial carbon dioxide emissions in Japan: an urban and institutional extension of the IPAT identity.** *Carbon Balance and Management* 2006, **1:** 11. 10.1186/1750-0680 1-11

13 Söderholm P, Ek K, Pettersson M: **Wind power development in Sweden: Global policies and local obstacles.***Renewable and Sustainable Energy Reviews* 2007, **11:** 365. 10.1016/j.rser.2005.03.001

14 Nguyen KQ: **Impacts of wind power generation and CO_2 emission constraints on the future choice of fuels and technologies in the power sector of Vietnam.** *Energy Policy* 2007, **35:** 2305. 10.1016/j.enpol.2006.06.023

15 Lewis JI, Wiser RH: **Fostering a renewable energy technology industry: An international comparison of wind industry policy support mechanisms.** *Energy Policy* 2007, **35:** 1844. 10.1016/ j.enpol. 2006.06.005

16 Ibenholt K: **Explaining learning curves for wind power.** *Energy Policy* 2002, **30:** 1181. 10.1016/S0301-4215(02)00014-9

17 Hays K, Attwood D: **Asia: Out of the shadows? Led by China and India, Asia Pacific wind markets poised for major growth.** *Refocus* 2006, **7:** 24. 10.1016/S1471-0846(06)70655-2

18 Carolin MM, Fernandez E: **Growth and future trends of wind energy in India.** *Renewable and Sustainable Energy Review* 2007.

19 Jagadeesh A: **Wind energy development in Tamil Nadu and Andhra Pradesh, India Institutional dynamics and barriers – A case study.** *Energy Policy* 2000, **28:** 157. 10.1016/S0301-4215(00)00007-0

20 Patel MR: *Wind and Solar Power Systems*. CRC Press, New York; 1999.

21 Burton T, Sharpe D, Jenkins N, Bossanyi E: *Wind Energy Handbook*. John Wiley & Sons, New York; 2001.

22 Spera DA: *Wind Turbine Technology: Fundamental Concepts of Wind Turbine Engineering. Volume Chapter 9*. ASME Press: New York; 1994.

23 [http://www.centreforenergy.com]

24 [http://www.windpowerindia.com]

25 MNES: **Annual Report: 2005–06.** In *Ministry of Non-conventional Energy Sources*. Govt. of India, New Delhi; 2006.

26 Ramachandra TV, Shruthi BV: **Wind energy potential mapping in Karnataka, India, using GIS.** *Energy Conversion and Management* 2005, **46:** 1561. 10.1016/j.enconman.2004.07.009

27 MNES: **Annual Report: 2004–05.** In *Ministry of Non-conventional Energy Sources*. Govt. of India, New Delhi; 2005.

28 Beurskens J, Jensen PH: **Economics of wind energy prospects and directions.** *Renewable Energy World* 2001, **4:** 103.

29 Nouni MR, Mullick SC, Kandpal TC: **Techno-economics of small wind electric generator projects for decentralized power supply in India.** *Energy Policy* 2007, **35:** 2491. 10.1016/j.enpol.2006.08.011

30 Amin AL: **Liberalization of the Indian power industry: Wind power in Gujarat.** *Renewable Energy* 1999, **16:** 977. 10.1016/S0960-1481(98)00345-0

31 Jagadeesh A: **Indian wind power: Wind as a supplementary energy source in India.** *Refocus* 2001, **2:** 16. 10.1016/S1471-0846(01)80043-3

32 [http://www.cd4cdm.org]

33 [http://cdm.unfccc.int/index.html]

34 UNFCCC: **Indicative simplified baseline and monitoring methodologies for selected small-scale project activity categories.** *Appendix B of the simplified modalities and procedures for small-scale CDM project activities. Bonn, Version 10* 23 December 2006. 23 December 2006.

35 [http://www.cea.nic.in]

36 Bode S, Michaelowa A: **Avoiding perverse effects of baseline and investment additionality determination in the case of renewable**

energy projects. *Energy Policy* 2003, **31:** 505. 10.1016/S0301-4215(02)00076-9

37 Shrestha RM, Timilsina GR: **The additionality criterion for identifying clean development mechanism projects under the Kyoto Protocol.** *Energy Policy* 2002, **30:** 73. 10.1016/S0301-4215(01)00054-4

38 Sugiyama T, Michaelowa A: **Reconciling the design of CDM with inborn paradox of additionality concept.** *Climate Policy* 2001, **1:** 75. 10.1016/S1469-3062(00)00005-X

39 TERI: **Cost of Unserved Energy.** In *TERI Project Report No. 98PG42*. Tata Energy Research Institute (TERI): New Delhi; 2001.

40 Bajaj: **Annual Report: 2000–2001.** *Pune* 2001, 18.

41 Gupta BK: **Weibull parameters for annual and monthly wind speed distributions for five locations in India.** *Solar Energy* 1986, **37:** 469. 10.1016/0038-092X(86)90039-3

42 Purohit P: **Financial evaluation of renewable energy technologies for irrigation water pumping in India.** *Energy Policy*2007, **35:** 3134. 10.1016/j.enpol.2006.11.013

43 Purohit P, Kandpal TC: **Techno-economic evaluation of water pumping windmills in India.** *International Journal of Global Energy Issues* 2004, **21:** 263. 10.1504/IJGEI.2004.005771

44 Islam MN, Haque MM: *Technology, planning and control.* World University Service Press, Dhaka; 1994.

45 Purohit P, Michaelowa A: **CDM potential of SPV lighting systems in India.** *Mitigation and Adaptation Strategies for Global Change* doi: 10.1007/s11027–006–9078-x.

46 Islam T, Meade N: **The diffusion of successive generations of a technology: a more general model.** *Technological Forecasting and Social Change* 1997, **56:** 49. 10.1016/S0040-1625(97)00030-9

47 Ang BW, Ng TT: **The use of growth curves in energy studies.** *Energy* 1992, **17:** 25. 10.1016/0360-5442(92)90030-4

48 Martino JP: **A review of selected recent advances in technological forecasting.** *Technological Forecasting and Social Change* 2003, **70:** 719. 10.1016/S0040-1625(02)00375-X

49 Purohit P, Kandpal TC: **Renewable energy technologies for irrigation water pumping in India: projected levels of dissemination, energy delivery and investment requirements using available diffusion**

models. *Renewable and Sustainable Energy Reviews* 2005,**9**(6):592–607. 10.1016/j.rser.2004.07.002

50 Purohit I, Purohit P: **CDM Potential of Box Type Solar Cookers in India.** *International Journal of Ambient Energy* 2007,**28:** 27.

51 Purohit P, Michaelowa A: **CDM Potential of Bagasse Cogeneration in India.** *Energy Policy* 2007, **35:** 4779. 10.1016/j.enpol.2007.03.029

52 Purohit P, Michaelowa A: **CDM potential of SPV pumps in India.** *Renewable and Sustainable Energy Reviews* 2006. doi:10.1016/j.rser.2006.05.011. doi:10.1016/j.rser.2006.05.011.

53 Purohit P, Michaelowa A: **CDM potential of solar energy technologies in India.** In *Renewable Energy 2006: Annual Magazine of World Renewable Energy Network (WREN).* Sovereign Press. London; 2006.

CHAPTER 6

Smart Grid for a Sustainable Future

G. M. Shafiullah*, Amanullah M. T. Oo, A. B. M. Shawkat Ali, Peter Wolfs

Higher Education Division, School of Engineering and Technology, Central Queensland University, Rockhampton, Australia.

ABSTRACT

Advances in micro-electro-mechanical systems (MEMS) and information communication technology (ICT) have facilitated the development of integrated electrical power systems for the future. A recent major issue is the need for a healthy and sustainable power transmission and distribution system that is smart, reliable and climate-friendly. Therefore, at the start of the 21st Century, Government, utilities and research communities are working jointly to develop an intelligent grid system, which is now known as a smart grid. Smart grid will provide highly consistent and reliable services, efficient energy management practices, smart metering integration, automation and precision decision support systems and self healing facilities. Smart grid will also bring benefits of seamless integration of renewable energy sources to the power networks. This paper focuses on the benefits and probable deployment issues of smart grid technology for a sustainable future both nationally and internationally. This paper also investigates the ongoing major research programs in Europe, America and Australia for smart grid and the associated enabling technologies. Finally, this study explores the prospects and characteristics of renewable energy sources with possible deployment integration issues to develop a clean energy smart grid technology for an intelligent power system.

KEYWORDS

Smart Grid; Renewable Energy; Grid Integration; Wind Energy; Solar Energy

1. INTRODUCTION

Existing power systems can be considered as a major reason for greenhouse or global warming effects that cause environmental impacts due to use of fossil fuels, especially coal. In contrast to fossil fuels, renewable energy (RE) offers alternative sources of energy which are in general pollution free, technologically effective and environmentally sustainable. There is unprecedented attention to RE, particularly solar and wind energy, which provides electricity without giving rise to carbon dioxide emissions. However, most of the current transmission and distribution networks are considered as "dumb" systems as they are not capable of feeding back the intelligent data required for a modern grid operation [1,2]. The existing electricity grid has no potential to offer adequate services addressing energy efficiency, reliability and security, or the integration of RE at the scale needed to meet the clean-energy demand for the future [3].

Therefore, introduction of smart grid technology is an essential requirement that reduces overall greenhouse gas (GHG) emissions with demand management that encourages energy efficiency, improves reliability and manages power more efficiently and effectively. Smart grid is the combination of centralized bulky power plants and distributed power generators that allows multi-directional power flow and information exchange. Its' two-way power communication systems create an automated and energy-efficient advanced energy delivery network. On the other hand, in traditional power systems, power flows only in one direction, i.e., from generating station to customers via transmission and distribution networks [1-6].

The smart grid is a broad collection of technology that delivers an electricity network with flexibility, accessibility, reliability and economy. Smart Grids are sophisticated; they can digitally enhance power systems where the use of modern communications and control technologies allows greater robustness, efficiency and flexibility than today's power systems [1-6]. Brief comparisons between an existing grid and a smart grid are given in **Table 1**.

Smart grid technologies are still new and many are in the development stage. However, it is anticipated that smart grid technology will be play a self-regulatory role in power system networks due to its many advantages as given below [6,7].

1.1. Intelligent and Efficient

Smart grid is capable of sensing system overloads and rerouting power to prevent or minimize a potential outage. It is efficient and potentially able to meet increasing consumer demand without adding any infrastructure.

1.2. Accommodating

Due to its robustness, smart grid can accommodate energy from fuel sources as well as RE sources and adopt any new technologies for a climate-friendly society.

1.3. Reduce Global Warming

Possible to integrate large-scale RE into the grid that reduces global warming as well as GHG emission.

1.4. Repairing and Maintenance

Automatic maintenance and operation increases the efficiency of the power network. Moreover, predictive maintenance and self-healing reduces system disturbances.

1.5. Reliability

Improves power quality and reliability as well as enhances capacity of existing network.

1.6. Distributed Generation

Accommodates distributed power sources efficiently which reduces energy costs, GHG emissions and energy crisis issues world-wide.

1.7. Consumer Focus

Consumers can customize their energy uses based on individual needs, electricity prices and environmental concerns.

1.8. Security

With the adoption of security features in smart grid, the network is safer from cyber-attack and any unwanted tampering and natural disaster.

1.9. Quality-Focused

Ensures power quality of the network by reducing voltage fluctuation (sag, swell and spikes) and harmonic effects in the network.

1.10. Technology

New concepts and technologies will be developed that enhance power system infrastructure and accommodate new opportunities in innovation.

1.11. Socio-Economic Development

This new technology will open new doors in the power sector and communication arena. It will play an active role in socio-economic development as well as create job opportunities.

Therefore, this study explores the research undertaken for the deployment of smart grid technology worldwide with their implementation challenges. The smart grid deployment issues that include: smart grid infrastructure, communication technologies, potential barriers, and development of possible viable solutions for such implementation were also explored. This study also aimed at developing an integrated platform to continuously investtigate the impacts of RE on the smart grid which will assist the power utilities to develop an improved national power grid that will help to build a sustainable society. The proposed integrated platform comprises with feasibility study to investigate the prospects of RE in Australian context; prediction model to assess the energy generation from RE sources; real-time experiments and simulation model to explore the impacts on integrating RE into the smart grid. This paper is organized as follows: Section 2 discusses the smart grid initiatives worldwide; smart grid deployment issues are represented in Section 3. Section 4 presents integration of renewable energy with smart grid. Section 5 concludes the paper with future directions.

Table 1. Comparison between existing grid and smart grid.

Existing grid	Smart grid
Mostly electromechanical	Digital in nature
One-way communication	Two-way communication
Mostly centralized generation	Distributed generation
Sensors are not widely used	Sensors are widely used
Lack of monitoring; only manual	Digital self-monitoring
Failures and blackouts	Adaptive and intelligent
Lack of control	Robust control technology
Less energy-efficient	Energy-efficient
Usually not possible to integrate RE	Possible to integrate large-scale RE
Customers have less scope to modify uses	Customers can check uses and modify

2. SMART GRID INITIATIVES WORLDWIDE

Substantial research, planning and development are required for the deployment of a new smart infrastructure to integrate large-scale RE into the energy mix. The European and North American vision for Smart Grid technology has evolved over the past decade and has begun to achieve a level of maturity. In the next two sub-sections, the international policy initiatives taken by Europe, the USA and Australia for smart grid developments for a sustainable future power system have been explored.

2.1. Smart Grid Developments in the USA and Europe

The USA's Center for American Progress imparts a view of a clean-electricity or clean energy "pipeline" which produces large scale renewable electricity, delivers electricity nationwide on a new high capacity grid, deals with all power generation and distribution with new robust information technology methods and allows consumers to contribute energy to the grid [2]. In April 2003, the Department of Energy (DOE), USA declared its "Grid 2030" mission, the vision of which was: Grid 2030 energizes a competitive North American Market place for electricity. It connects everyone to abundant, affordable, clean, efficient and reliable electric power anytime, anywhere. It provides the best and most secure electric services available in the world [8].

To fulfill this vision, Electrical Power Research Institute's (EPRI's) Intelligrid^{SM} has undertaken an initiative to develop the technological foundation for a smart power grid that links electricity systems with communication and computer technology to achieve tremendous gains in reliability and customer services [9]. To achieve this, Intelligrid architecture offers a methodology, tools and recommendations for integrating data networks and equipment that enable interoperability between products and systems [10]. **Figure 1** shows a schematic diagram of EPRI's Intelligrid model.

The US Department of Energy Office of Electric Transmission and Distribution's GridWise^{TM}program [11] is an industry alliance with a vision for the future electric system built upon the primary principle that information technology will intensely transform the planning and operation of the power grid. The GridWise program has a focus on:

- Communications architecture and standards
- Simulation and analysis tools
- Smart technologies
- Test beds and demonstrated projects
- New regulatory, institutional and market frameworks The impact of renewable energy and smart grid interactions were investigated by National Grid, USA [12] on the US power networks. A typical scenario of their proposed model is given in **Figure 2**.

The European Union Advisory Council established Smart Grids European Technology Platform in 2005, to develop the weak grid with smart grid technology and overcome the drawbacks in the existing power systems. The smart grid European Technology Platform [13] vision is that Europe's electricity network must be flexible (fulfilling customer needs), accessible (access to all network users, particularly for RE sources and high efficiency local generation with low carbon emission), reliable (assuring security and quality of supply) and economic (cost and energy efficient management) to facilitate Europe's electricity grids to meet the challenges and opportunities of the 21st century and fulfill the expectations of society. The key requirements to achieve the vision include [13]:

- Creating technical solutions to be applied promptly and cost effectively that allow grids to accept power additions from all energy sources.
- Harmonizing regulatory and commercial structures in Europe to facilitate cross-border dealing of both power and grid services, ensuring that they will accommodate a wide range of operating conditions.
- Launching technical standards and protocols that will ensure open access.
- Developing information, computing and telecommunication systems.
- Ensuring the successful interfacing of new and old equipment designs.

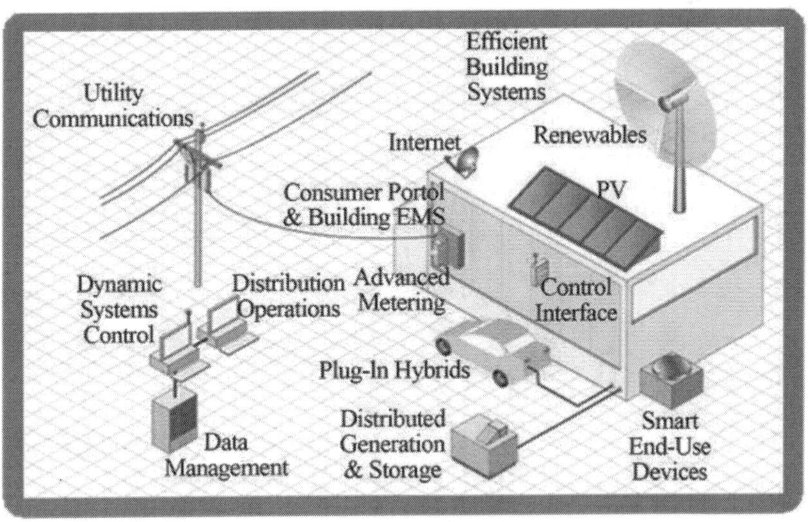

Figure 1. A typical scenario of EPRI's intelligrid model [9].

The future network vision of the European Technology Platform is shown in **Figure 3**.

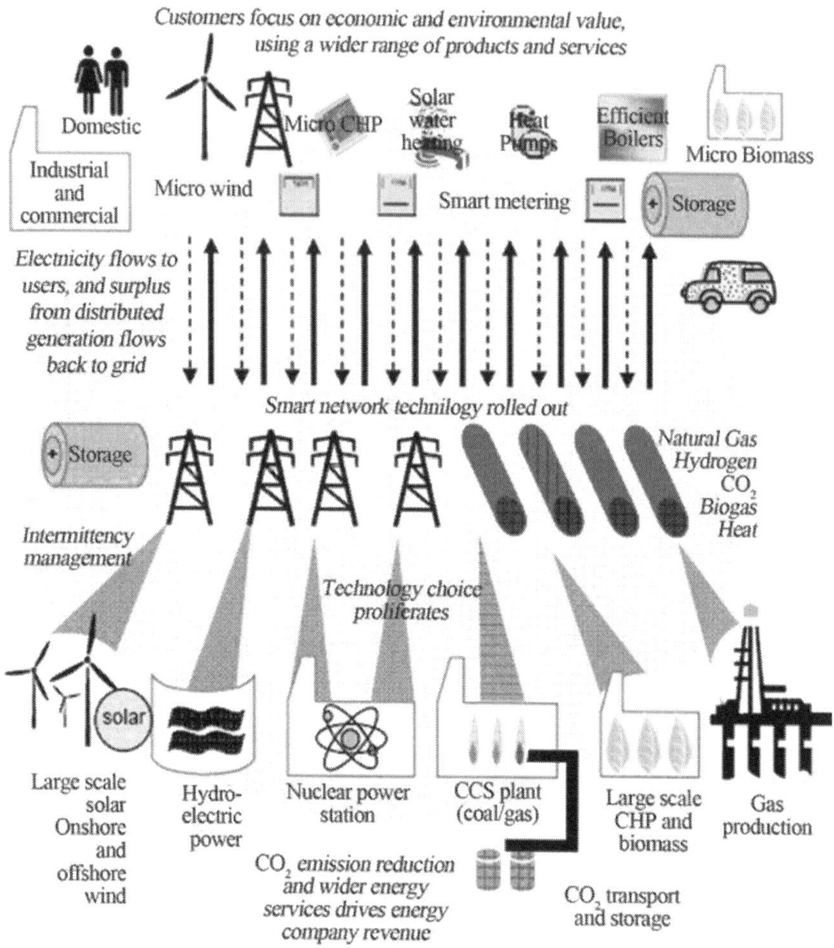

Figure 2. The smart grid evolution—A proposed model by national grid, USA [12].

2.2. Smart Grid Developments around Australia

Australia's energy prices are comparatively low as more than three-quarters of the country's electricity is generated by coal-fired plants. Hence, Australia is lagging behind compared to the USA and Europe in an attempt to integrate RE sources and build smart grid infrastructure. However, Australia's reliance on coal-fired power gives it one of the world's highest per-capita GHG emission rates. Coal imposes environmental costs in the form of greenhouse gases, including 200 million tons of carbon dioxide equivalents (CO_2-e) released in 2008, more than a third of Australia's total CO_2-e emissions [1,2]. Therefore, it is essential to strengthen the Australian economy by fostering new generation transmission and distribution systems for efficient use of low-carbon electricity, as the current high-voltage transmission grid imposes significant constraints on the deployment of new RE such as wind, solar and geothermal power. Current monitoring and control technology on both transmission and distribution networks is fragile [2]. The lack of smart technology to provide utilities and consumers with better information in real time impair the security and capability of the entire electricity system [1,2]. Therefore, it is necessary to integrate information processing and communications into power systems to create a unified smart grid that includes generation, transmission, distribution, retail and end-use.

The Australian Government has already taken the initiative with huge scale investment to develop their electricity infrastructure as well as deploy smart grid technology to integrate large scale RE into the grid [14]. Table 2 shows the current and future investments in Australia by State in which the green bars depict current periods and red bars future periods. Recent determinations for the ACT and NSW increase expenditure allowances by 70% over the preceding funding period [14]. Similar increasing trends are also observed in other States.

The Australian government has adopted the Smart Grid, Smart City initiative and invested $100 million to create a large scale smart grid platform, which optimizes societal benefits by prioritizing applications and undertaking a commercial scale deployment that tests the business case and main technologies [1]. It investigates synergies with other Australian Government programs and related infrastructure projects, builds awareness of the benefits and obtains buy-in from industry and customers. This project seeks to overcome major hurdles by supporting the adoption of smart grid technologies across Australia and build an innovative solution [1]. The aim is to link 9500 homes to a smart grid to allow residents to see real-time electricity usage in their houses which helps them to optimize the use of electricity. The Energy Australia Consortium is working hard to complete the project within the allocated timeframe and has already connected a large number of households with significant telecoms backbone infrastructure. The project will also evaluate smart grid applications in integrating RE and electric vehicles. The project is expected to be completed by 2013 [15,16].

In Australia, the Intelligent Grid Program [17] was commenced on 19 August 2008, being recognized under the CSIRO's Energy Transformed Flagship. The main focus of the program is to reduce GHG emissions [1]. The Townsville Solar City Project managed by the Australian Government and Ergon Energy (a local Queensland based electricity distribution utility organization) has conducted 742 residential and commercial assessments, and installed 1445 smart meters, 160 kW of solar panels and eight advanced energy storage systems. A range of schemes that aim to cut down wasteful energy usage, increase solar energy usage and reduce GHG emissions by more than 50,000 tonnes has been undertaken in the project [18]. Western

Power's (a local Western Australia based transmission and distribution utility organization) Solar City program includes a PV saturation trial to test the effect of distributed generation on the network. The Solar cities program has assisted many distributors to realize the problems of inverter connected renewable distributed generation (DG). This analysis included smart meters that collect bi-directional data to capture how much power the distributed generator is feeding back to the grid [1,19].

Figure 3. Future network vision [13].

Table 2. Investments by state: current and future [14].

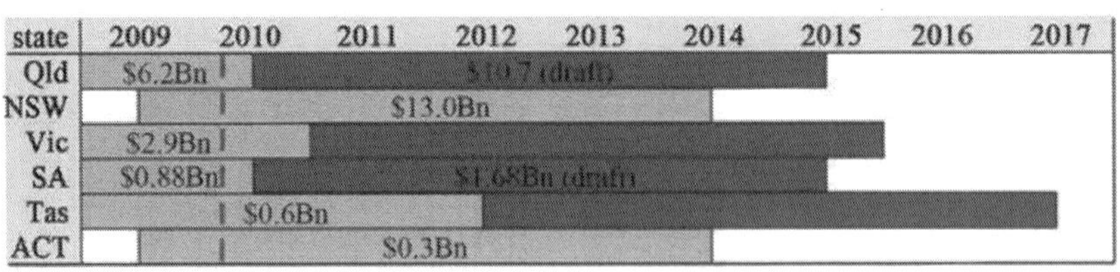

Recently, the Australian government has adopted the Clean Energy Initiative (CEI) for the operation of a range of renewable and clean energy technologies to reduce GHG emissions and meet rising electricity demand. The CEI program consists of Carbon Capture and Storage (CCS) initiatives, Solar flagships, the Australian Solar Institute (ASI), the Australian Centre for RE (ACRE) and the RE Future Fund (REFF) [20].

3. SMART GRID DEPLOYMENT ISSUES

With the initiative of international policy makers and Governments, research communities, industries, and scientists are working together to develop the smart grid and its enabling technologies for a robust power system worldwide. In this section, three major systems required to deploy smart grid technologies in the power system networks are explored, comprising the infrastructure of smart grid technologies, the associated communication technologies and standards and, finally, a smart grid model for the future.

3.1. Smart Grid Infrastructure

The smart grid is capable of integrating RE into the grid and delivering power in more efficient ways by utilizing modern information technologies as well as smart infrastructure and control management systems. Fang et al. [5] conducted an extensive survey in which the smart grid and its enabling technologies were divided into three major sections:

3.1.1. Smart Infrastructure System

Smart infrastructure system comprises the smart energy subsystem, the smart information subsystem and the smart communication subsystem. Electricity generation, transmission and distribution, as well as consumption of electricity facilities, are incorporated in the smart energy subsystem. The smart information subsystem comprises smart metering, advanced monitoring and management of the smart grid network. The smart communication subsystem is responsible for wired and wireless communication between networks, devices and applications to established connectivity in the network [5].

3.1.2. Smart Management System

Advanced management, monitoring and control services are provided by the smart management system in smart grid. With the development of advanced management and monitoring applications and control services, the smart grid technology will be smarter and play an active role in developing a sustainable power system. The smart management system comprises energy efficiency improvement, supply and demand balance, emission control, operation cost reduction and utility maximization. Modern machine learning and other optimization tools are used to make a robust smart management system [5].

3.1.3. Smart Protection System

Reliability, failure protection, security and privacy protection services of the grid are provided by the smart protection system in smart grid. Integration of advanced protection equipment and monitoring tools ensures network reliability, security and privacy. In addition to smart infrastructure design, smart management and smart protection systems address efficient management and failure protection, cyber security and privacy in the network. **Figure 4** shows a representation of the typical technological architecture of the smart grid [5].

3.2. Smart Grid Communication Technologies and Standards

Smart grid technologies are basically used for smart power delivery which is a combination of current and advanced new technologies. Smart grid technologies include automation technologies for smart power delivery, new advanced communication technologies, distributed energy and storage technologies such as solar, wind turbines and fuel cells, advanced metering infrastructure (AMI), power electronics-based controllers and appliances and devices which are demand-response ready [3]. Smart grid has enhanced and robust communication and computing capabilities that makes this an attractive technology for future power systems.

Smart grid ensures today's key requirement of reliable delivery of power from the generating stations to end-users by introducing modern information and communication standards and technologies [21]. Recent studies [21-23] show that both wired and wireless communication systems are used in smart grid for data communication between smart meters and electric utilities. The data flow from sensors and electrical appliances to smart meters can be accomplished by power line communication or wireless communication. Cellular technologies or the internet can be used to communicate between smart meters and a utility's data centre. A typical smart grid architecture proposed by Gungor et al. [21] is presented in **Figure 5**.

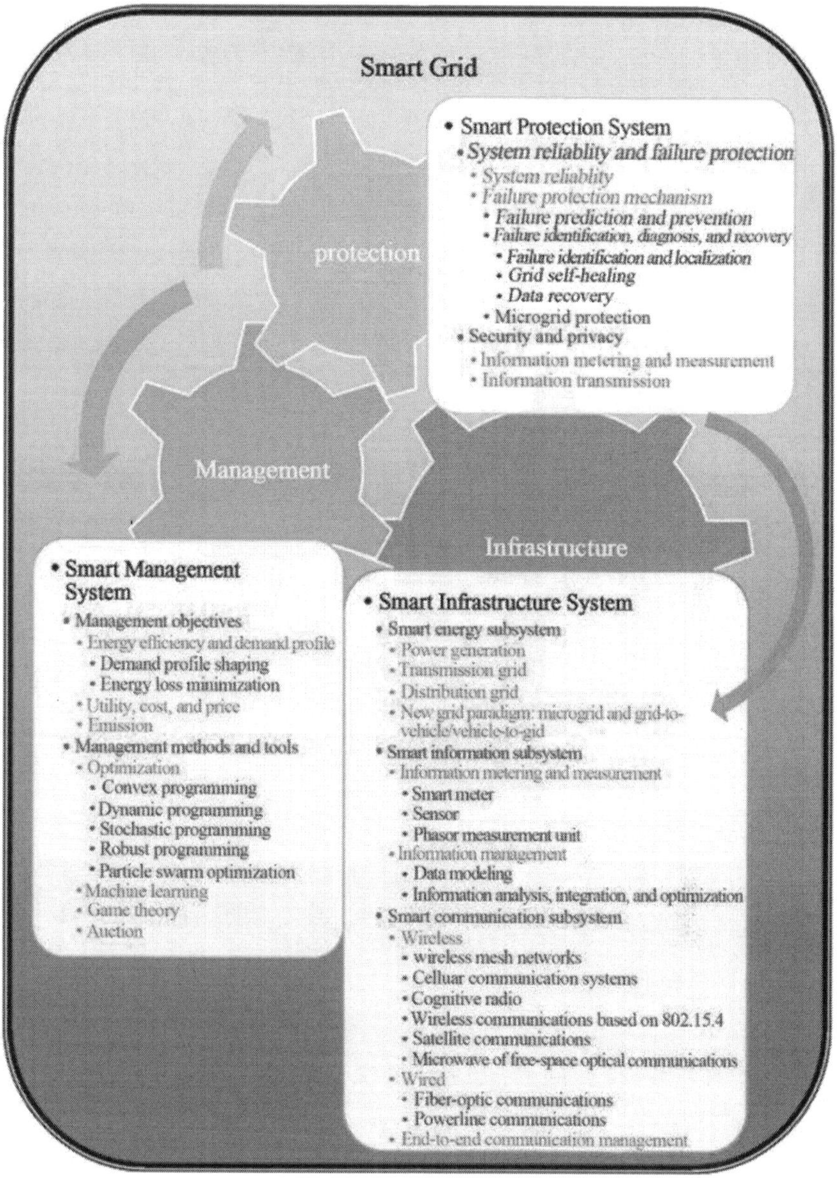

Figure 4. The typical classification of the smart infrastructure, the smart management and the smart protection system [5].

Popular communication technologies that are used in smart grid are ZigBee, wireless mesh, cellular network communication, power line communication (PLC) and digital subscriber lines (DSL) though they all have some advantages and disadvantages. Wired technologies, such as PLC, DSL and optical fiber are reliable and secured, however, they require high capital cost due to large-scale installation and maintenance. On the other hand, wireless technologies require less installation cost though they have problems with bandwidth and security [21]. Ensuring reliability and security in the communication medium or path is essential for appropriate end-to-end data communication. Smart grid should be secure enough to prevent cyber-attacks and have appropriate bandwidth allocation with an advanced control system that provides stability and availability. Wireless technologies with controlled bandwidth and security are a preferable option as it is possible to deploy large scale smart grid with reduced costs, while wired technologies with high capacity and security require high costs [24]. However, considering costs, robustness and reliability, a hybrid communication technology combining both wired and wireless techniques has been used.

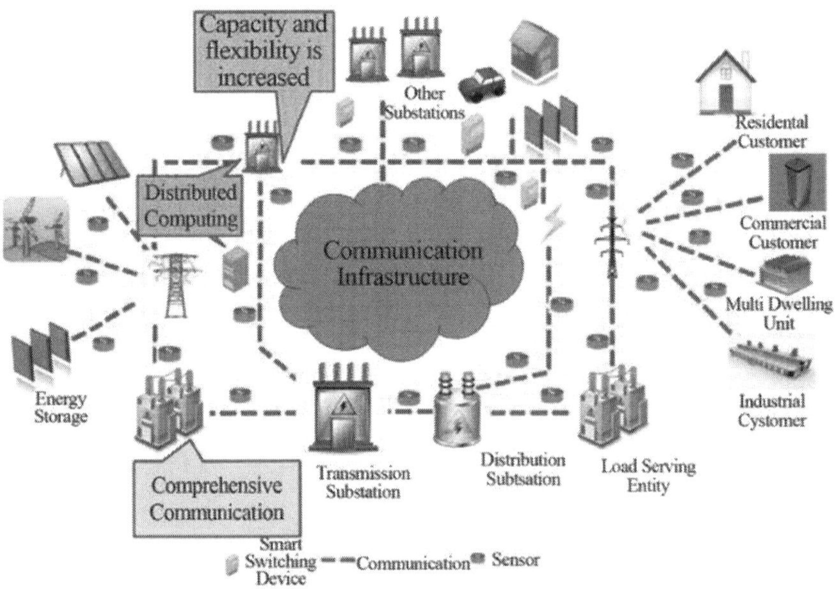

Figure 5. Smart grid network with modern communication technologies [21].

Researchers are working together to investigate new communication technologies, applications and standards for a secure, reliable, robust and efficient smart grid network. The goal of smart grid standards is to provide seamless interoperability between systems, information security, a compact set of protocols and communication exchange [25]. The European Technology Platform, The American National Standards Institute, the International Electrotechnical Commission, The Institute of Electrical and Electronics Engineers, the International Organization for Standardization, the International Telecommunication Union and the Third Generation Partnership Project are the recognized standard development organizations working together and focused on smart metering functionalities and communication interfaces.

3.3. Smart Grid Model

Recently, Zahedi [3] proposed a smart grid model that is useful in understanding the architecture and power flow of the smart grid as shown in **Figure 6**. The major parts of the model are:

3.3.1. Central Power Stations

Power is generated from various energy sources in the central power stations. The sources of energy are renewable such as solar, wind and hydro, and non-renewable such as coal-fired, nuclear and gas turbine, etc. Suitable storage facilities are also included in the power stations.

3.3.2. Transmission Network

Bulk high-voltage electricity passes through transmission lines over long distances in the smart grid. Smart equipment devices are used to transfer electricity from transmission substations to distribution substations.

3.3.3. Distribution Network

The distribution network model delivers electricity to end-users at low-voltage levels. In the distribution network, there are bi-directional power flows as the system absorbs energy from the consumers who generated power through roof-top solar panels, etc. The distribution network connects smart meters and all intelligent devices and controls them through a two-way wireless and wired communication network [2,26].

3.3.4. Market Domain

In the model, the market domain both comprises and coordinates all the participants in electricity markets serviced by the smart grid.

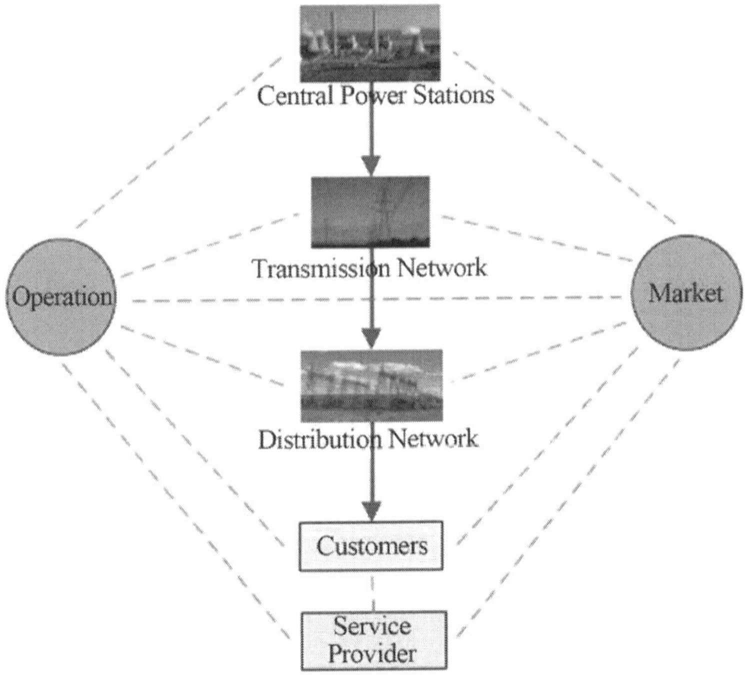

Figure 6. Typical smart grid model [3].

3.3.5. Service Provider Domain

Service providers supply services to both customers and power utilities.

3.3.7. Operation Domain

The operation domain maintains and controls the electricity flow of the other domains within the smart grid.

3.3.8. Customer Domain

The customer domain connects the residential, commercial and industrial customers to the electric distribution network through smart meters.

This model is expected to be used as the basis for the design of smart grid infrastructure that defines characteristics, requirements, interfaces and performance of the grid [3].

Therefore, it can be concluded that International policy makers, Government, industrialists, and researchers are working together to deploy smart grid technology worldwide for a sustainable robust and secured power systems for the future. The utilization of smart grid functions allows for larger saturation of variable energy sources through more adaptable management of the system [27]. However, there are a number of challenges in integrating RE sources with the smart grid enabled power systems.

Therefore, the authors are currently investigating the characteristics and prospects of RE sources, as well as impacts on integrating RE with the smart grid, to develop a clean-energy smart grid technology for a sustainable future both nationally and internationally. A snapshot of the work in progress is presented in the next section.

4. INTEGRATION OF RE WITH SMART GRID

The advances of smart grid technologies have guided the growth of information technology based energy management and climate change mitigation systems. The authors are currently investigating the benefits and probable mixing of RE into the smart grid so as to build up a clean energy system for a sustainable future. This project looks at the strategic impacts of integrating RE sources with the grid together with the analysis of the possible limitations, and developments of probable feasible solutions, for such implementation. An integrated platform will be developed to continuously investigate the effects of RE on the smart grid which will support the power utilities in developing an improved national power grid.

To investigate the impact of RE sources within a smart grid, the methodology and approach to carry out the research includes: investigation of existing transmission and distribution networks; conducting a feasibility study on integrating RE sources with the existing distribution networks; real-time experiments using RE sources; investigation of the potential challenges of RE inputs to a smart power grid; and finally, development of a tool or platform to continuously investigate the impacts of RE on smart grids. **Figure 7** presents the framework of the proposed study. In the following subsections, details of each approach are presented.

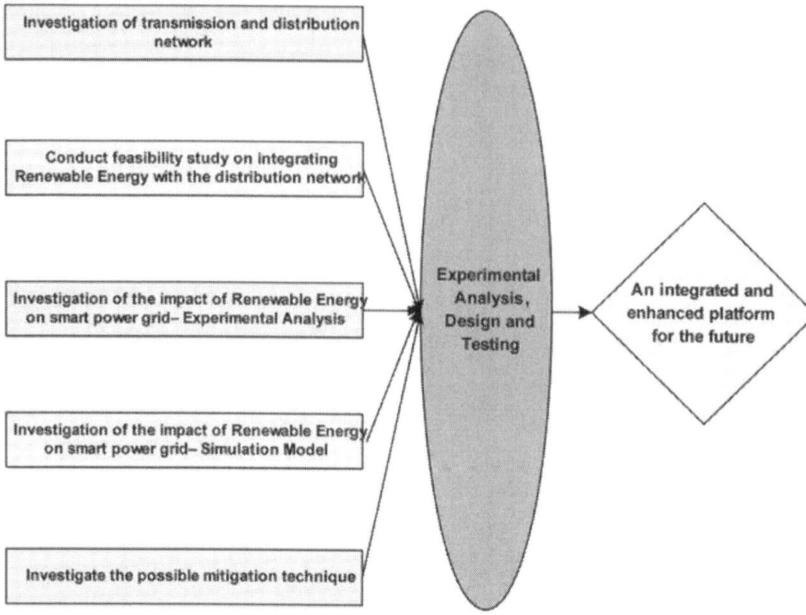

Figure 7. Block diagram of the planned methodology framework.

4.1. Prospects of RE in the Australian Context

Hybrid renewable energy models were developed to investigate the prospects of renewable energy, in particular wind and solar energy. Suitable locations for both wind and solar energy generation were investigated using statistical analysis. To assess the performances of different hybrid systems in the study, optimal systems and the sensitivity analysis were undertaken using the HOMER simulation tool [28]. The models were simulated based on the sensitivity variables of wind speed, solar irradiation and grid electricity price. Net present cost (NPC), cost of energy (COE) and renewable fraction (RF) [29] were estimated as model outputs to evaluate the performances of the studied models. PV/wind/grid-connected, PV/grid-connected, wind/grid-connected and only grid-connected systems were evaluated to analyze the potentialities of RE in Australia. The details of the study are available in Ref. [30].

From the statistical analysis it was estimated that, for wind energy generation, Tasmania ranked 1 (most suitable) and the Northern Territory ranked 7 (least suitable) out of the seven States of Australia. On the other hand, it was estimated that the Northern Territory ranked 1 and Tasmania ranked 7 out of the seven States of Australia for solar energy generation.

From simulation analysis, it was seen that the wind turbine and PV module required almost 80% of the total cost as capital expenditure. However, most of the operating cost is required for the grid components and converters, and RE sources require less operating expenditure which is one of their most useful features.

It was observed that Flinders Reef, Queensland, Australia is the most suitable place considering the energy generation from both solar and wind resources, with a contribution from RE of 90% as shown in **Figure 8**. From that Figure, it can be seen that a wind/grid-connected system is more suitable than the PV/wind/gridconnected system. It is also seen that a wind/grid-connected power system is economically feasible with a minimum total NPC of $75,789 and a minimum COE of $0.162/kWh, while a grid-connected only power system requires a minimum total NPC of $184,055 and a minimum COE of $0.394/kWh.

This feasibility study evaluated the total yearly emissions from PV/wind/grid-connected, standard grid-connected, wind/grid-connected, and PV/grid-connected systems. **Figure 9** presents the yearly emissions of CO_2, SO_2, and NO_2 for the developed hybrid RE model at seven selected locations in Australia. For example, for Ceduna in South Australia, the standard grid-connected only system emitted 34,018 kg CO_2 per year; on the other hand, a PV/wind/grid-connected system emitted only 10,842 kg/yr. A PV/grid-connected system emitted 32,794 kg CO_2 per year and a wind/grid-connected system emitted only 11,773 kg/yr. **Figure 9** shows that a small number of places have negative emissions especially for wind/ gridand PV/wind/grid-connected systems; i.e., these systems sell more power than they purchase from the grid as they generate surplus electricity [30].

Therefore, it can be concluded that large-scale penetration of RE not only reduces the cost of energy and assists with the energy-crisis worldwide, but also reduces emissions which will have a major role in building a sustainable climate-friendly future both nationally and internationally. However, it was found that wind energy has more potentialities compared to solar energy.

Based on the performance metrics, optimization and sensitivity analysis, it was observed that most of the selected locations have prospects for solar and wind energy generation. However, few locations perform exceptionally well for wind energy generation and few locations are greatly capable for solar energy generation.

4.2. Prediction Model of RE

To facilitate the introduction of a large number of RE sources into the grid, a careful evaluation of the attributes of each RE source is crucial, in particular the variability of its production with changing weather conditions. The useful utilization of wind and solar energy requires a detailed knowledge of the wind and solar features at the particular location, and the distribution of wind speeds and solar irradiation is important for the design of wind farms and solar plants, and power generators. However, adequate information is not always available over reasonable periods of time on the wind power or solar plant sites for power system planning purposes. Therefore, prediction models were developed that estimates the typical power generation available separately from wind and solar sources in advance using modern machine learning techniques.

This study was conducted using ten different regression algorithms to forecast the daily distribution of solar irradiation and hourly distribution of wind flow. Metabased learning Random Sub Space (RSS), and RegressionByDiscretization(RegDes), Regression-based learning linear regression (LR), simple linear regression (SLR), Statistical learning based algorithm sequential minimal optimisation (SMO) regression (SMOReg), and Neural Network based multilayer perception (MLP), RBFNetwork (RBFN), Lazy-based learning IBK, Treebased learning M5Rules and RepTree with bagging techniques [31-33] were considered in this study for developing the prediction model. The most suitable algorithm was proposed based on the performance metrics [31] of the algorithms that include the correlation coefficient (CC), mean absolute error (MAE), root mean square error (RMSE) and computational complexity. The Waikato Environment for Knowledge Analysis (WEKA) learning tool [34] was used to evaluate the prediction accuracy. Proposed algorithms with classical data splitting options were used to predict the daily distribution of solar radiation and hourly distribution of wind speed. Popular statistical analyses were carried out to select the most suitable algorithm to predict solar irradiation and wind speed as well as energy generation from these resources. Details of the model are available in Reference [35].

Initially, models were developed using ten regression algorithms with the collected hourly wind data to predict hourly distribution of wind flow. Model results showed that overall prediction accuracy is fairly similar; however no algorithm performs the best for all of the estimated attributes. **Figure 10** shows that IBK, RBFN and RegDes predicted with highest accuracy. CC is the least accurate for the model developed with linear regression (LR). In terms of MAE and RMSE, the model developed with RepTree with Bagging techniques is the best.

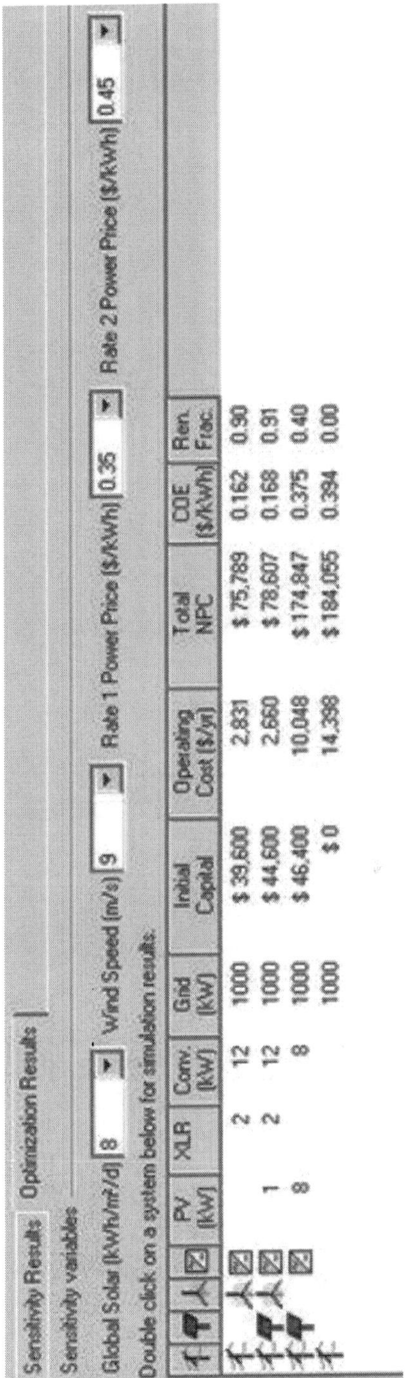

PV (kW)	XLR	Conv. (kW)	Grid (kW)	Initial Capital	Operating Cost ($/yr)	Total NPC	COE ($/kWh)	Ren. Frac.
	2	12	1000	$ 39,600	2,831	$ 75,789	0.162	0.90
1	2	12	1000	$ 44,600	2,650	$ 78,607	0.168	0.91
8		8	1000	$ 46,400	10,048	$ 174,847	0.375	0.40
			1000	$ 0	14,398	$ 184,055	0.394	0.00

Figure 8. Optimization results for flinders reef, QLD with 8.0 kWh/m²/d solar radiation, 9.0 m/s wind speed [30].

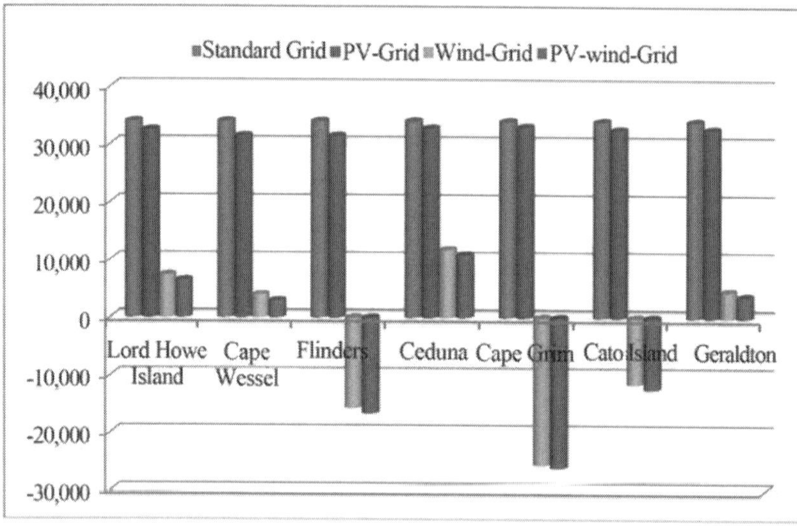

Figure 9. CO_2 emissions at seven selected locations for the four systems studied [30].

Rankings for the algorithms have been estimated and it is observed that Bagging technique has ranked 1, while RBFN has ranked 2 and RegDes has ranked 10. Finally, relative weighted performance was measured considering average ranking accuracy and computational complexity using the following equation:

$$Z = \alpha a_i + \beta t_i \tag{1}$$

here, α and β are the weight parameters for ranking average accuracy against computational complexity. The average accuracy and computational complexity are denoted by a_i and t_i respectively. From **Figure 11** it is shown that, considering computational complexity and average accuracy, Bagging technique is the best choice for all β values and RBFN is the second for this application. The model developed with RegDes performs the worst for this application. It has also been observed that MLP requires more computational time compared to other algorithms.

Similar models have been developed using the same ten regression algorithms with solar irradiation data. From the results, it was seen that the models developed with RSS and Bagging technique with RepTree are respectively the first and second prefered to predict daily solar irriadiation as well as solar energy production.

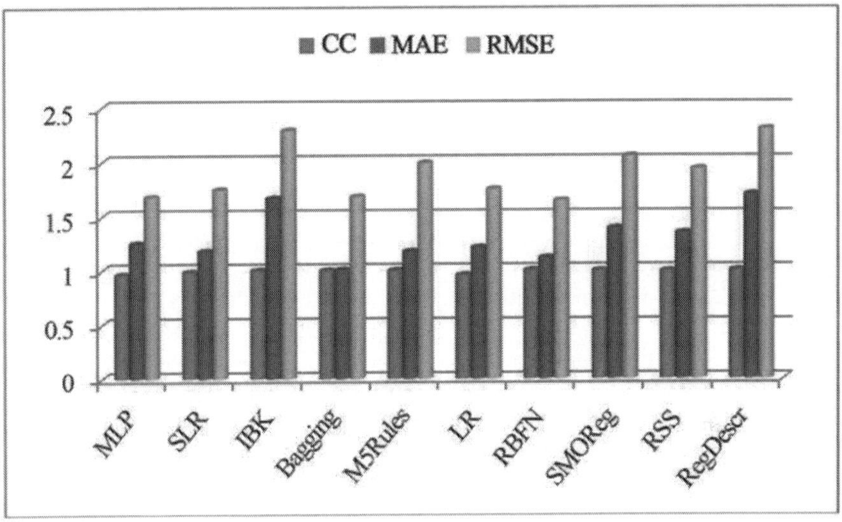

Figure 10. Comparisons of performance metrics with different algorithms for prediction of houlry wind speed [35].

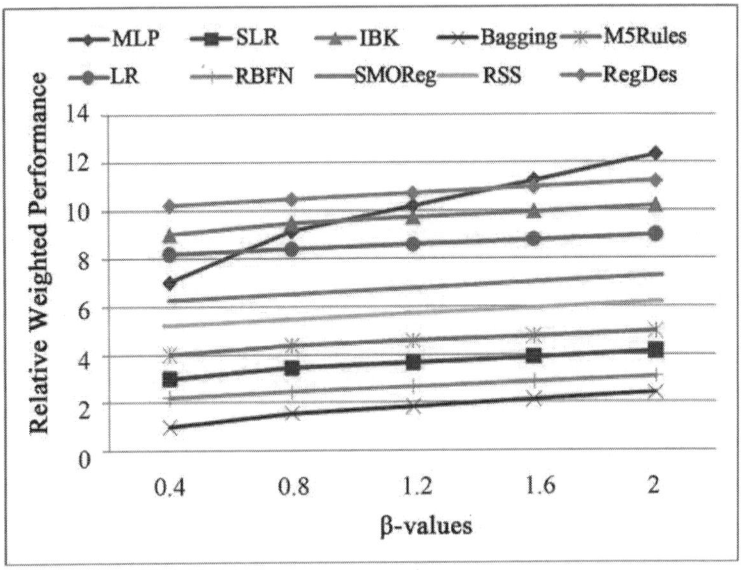

Figure 11. Overall weighted performance of the algorithms with respect to β for wind speed data [35].

4.3. Impacts on Integrating RE into the Grid

To analyze the impacts of RE integration, in particular the impacts of large-scale PV integration into the grid, experiments were undertaken at CSIRO, Newcastle, Australia [36] using their Renewable Energy Integration Facility (REIF). Voltage instability and harmonic impact on the network was explored with varying PV saturation and load conditions. From this work, it was possible to measure voltage instability, harmonic effects, reactive and real power with the increases of PV saturation and increases of micro turbine capacity with varying load conditions. A detail of the study is presented in Reference [37].

Considering the flexibility and robustness of the experiments, two experiments were carried out with different case scenarios. Experiment 1 was performed with 11.31 kW·PV saturation with varying micro turbine and load conditions. On the other hand, Experiment 2 was carried out with 7.5 kW·PV saturation with varying micro turbine and load conditions. From the experimental results, it was show that the harmonic content of the network increases with the increase of PV saturation and system size. Experimental results also show that significant voltage instability occurs into the system and the power factor is reduced with increasing PV penetration.

Neutral currents for 11.31 kV and 7.5 kV·PV saturation are shown in **Figure 12**. From that Figure, it can be seen that the neutral current fluctuates significantly from the original sinusoidal wave which indicates that the system is unbalanced and introduces a large amount of harmonics into the system. It has also been found that harmonics injection increases with increasing PV saturation.

In this study, Fast Fournier Transform (FFT) was applied to investigate harmonic effects in the distribution system. Based on the Australian standard, 50 Hz is taken as the fundamental frequency and the harmonic contents were located at the integer multiples of fundamental frequency. From the spectral analysis, it was found that, with the saturation of PV, current harmonics are introduced into the system. It was shown that, other than the fundamental frequency, maximum current is at 450 Hz, i.e., in the 9th harmonic. Levels of individual harmonics are shown in **Figure 13**, from which it can be concluded that harmonic introductions from all even harmonics are within the safety boundary as stated in the AS4777 standard. However, introductions from 3rd and 9th harmonics surpass the regulatory standard, while introductions from 7th and 15th harmonics just touch the threshold levels. However, harmonic current in the network would be considerably increased with PV saturation.

Moreover, from the experiments it was shown that, for the variation of PV generation and inverter characteristics, minor voltage instability as well as reduction in power factor occur in the network. However, to properly examine these problems, large-scale PV saturation into the network is required. Hence, this study will further examine the potential barriers to large-scale PV saturation into the grid with simulation analysis.

The authors are now developing a simulation model using PSS SINCAL [38] that mimics the experimental setup to evaluate the effects of PV saturation into the grid. Preliminary results from simulation show that voltage instability and harmonic introduction increases with increasing PV saturation. In some cases, these values exceed the regulatory standards.

The final contribution of this study is to develop an integrated platform which will enable researchers and Industry practitioners to effectively and efficiently investigate the impacts of the integration of RE into the future smart grid power system. This proposed tool will be developed by integrating all of the developed modules discussed throughout this study into an interactive user interface.

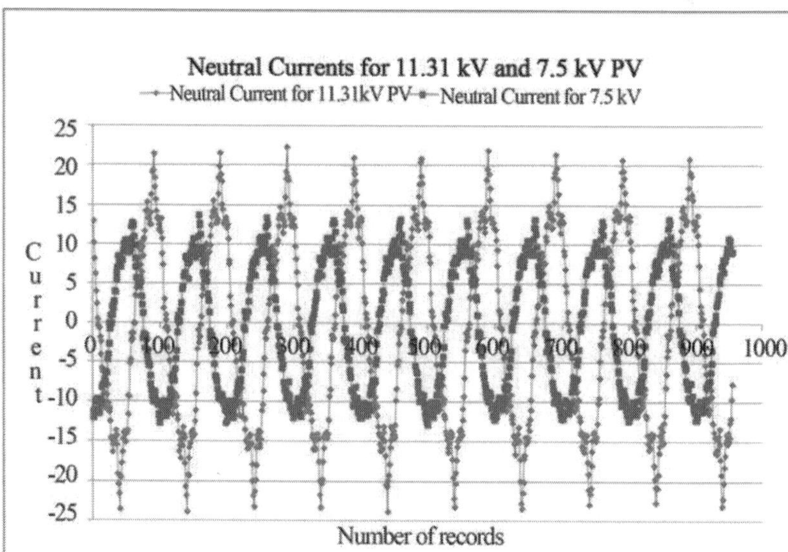

Figure 12. Neutral currents with 11.31 kV and 7.5 PV [37].

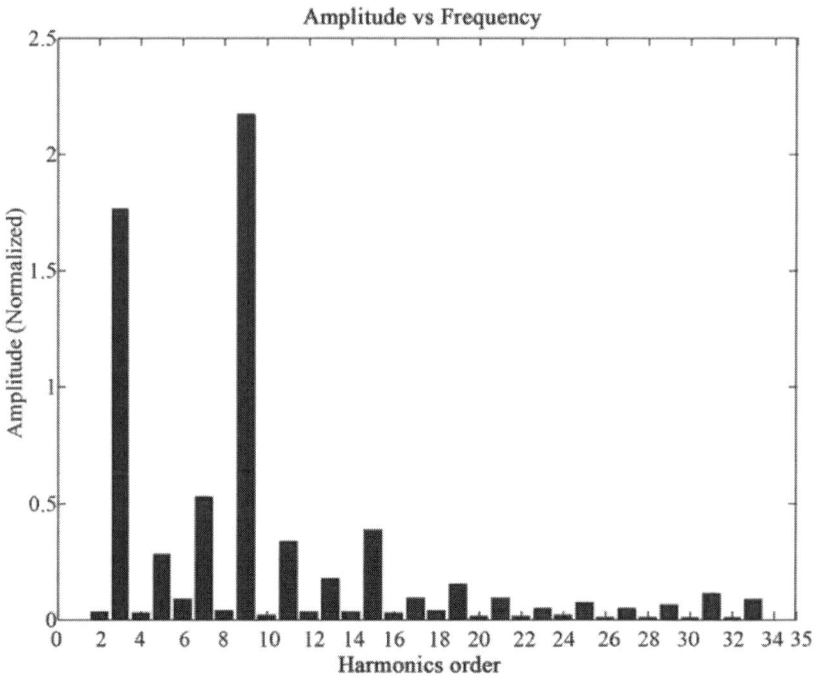

Figure 13. Amplitudes of different harmonics [37].

5. CONCLUSION

Recent environmental consciousness resulting from conventional power station has encouraged attention on the development of modern smart grid technology and its integration with climate friendly green renewable energy. The use of smart grid operations allows for greater penetration of variable energy sources through more flexible management of the system. Considering the current scenario, substantial research, planning and development work has been undertaken world-wide, especially in Europe, the USA and Australia. The USA declares its mission of “Grid 2030” that connects everyone to an abundant, affordable, clean, efficient and reliable electric power network. The primary national goal of Australia is to increase the use of renewable energy from present levels to 20 percent of total electrical supply by 2020.

As a work in progress, in this paper a feasibility study was investigated exploring the characteristics and cost analysis of grid connected hybrid RE systems, in which it was clearly observed that RE has significant potentialities. Experiments were carried out to investigate the impacts of RE sources in a smart power network, in particular large-scale PV penetration into the grid. From experimental and simulation analyses, it was observed that voltage fluctuations and harmonic injection increases with the increase of PV penetration. However, this study is still in the introductory stage and needs to be further investigations on the following areas to successfully complete the study:

- Analyze the impacts of large-scale integration of RE sources into the smart power systems.
- Develop integrated platform to monitor continuously the impacts of RE in developing a smart power system for the future.

REFERENCES

1 "Smart Grid, Smart City: A New Direction for a New Energy Era," Technical Report, Department of the Environment, Water, Heritage and the Arts, 2009.

2 B. Hendricks, "Wired for Progress—Building a National Clean-Energy Smart Grid," Technical Report, Center for American Progress, 2009.

3 A. Zahedi, "Developing a System Model for Future Smart Grid," Proceedings in 2011 IEEE PES Innovative Smart Grid Technologies Conference, ISGT Asia 2011, Perth, 13-16 November 2011, pp. 1-5.

4 "Proceedings and Outputs of the Workshop for Developing the Australian Smart Grid R&D Roadmap," Technical Report: SGA Research Working Group and CSIRO, Sydney, 2009.

5 X. Fang, S. Misra, G. Xue and D. Yang, "Smart Grid— The New and Improved Power Grid: A Survey," International Journal of IEEE Communications Surveys & Tutorials, Vol. 14, No. 4, 2011, pp. 944-980.

6 B. Panajotovic, M. Jankovic and B. Odadzic, "ICT and Smart Grid," Proceedings for the 10th International Conference on Telecommunication in Modern Satellite Cable and Broadcasting Services, Serbia, 5-8 October 2011, pp. 118-121.

7 "The Smart Grid: An Introduction," Technical report by US depart of Energy, 2008. http://energy.gov/oe/downloads/smart-grid-introduction-0

8 "Grid 2030—A National Vision for Electricity's Second 100 Years," US Department of Energy, Office of Electric Transmission and Distribution, July 2003.

9 "Electric Power Research Institute Intelligrid," Technical Report. http://intelligrid.epri.com/

10 R. E. Brown, "Impact of Smart Grid on Distribution System Design," Proceedings of IEEE Power and Energy Society General Meeting—Conversion and Delivery of Electrical Energy in the 21st Century, Pittsburg, 20-24 July 2008, pp. 1-4.

11 GridWise Program, "Rethinking Energy from Generation to Consumption," US Department of Energy, Office of Electric Transmission and Distribution, January 2005.

12 B. J. Walker, "Renewable Energy & Smart Grid Interactions," Technical Report: National Grid—The Power of Action, New England Conference of Public Utilities Commissioners, 2009.

13 "Smart Grids—European Technology Platform," Technical Report. http://www.smartgrids.eu/documents/vision.pdf

14 M. Davidson, "Smart Grid Australia: An Overview," Technical Report: Wessex Consult, Australia, 2010.

15 P. Budde, "Smart Grid Australia Provided the Vision for the Smart Grid/Smart City Project," Smart Grid Australia, October 2010.

16 "Smart Grid, Smart City," Report: Department of Resources, Energy and Tourism, Australian Government, May 2011.

17 "Grid Intelligent Grid," Technical Report, http://www.igrid.net.au/

18 "Townsville Solar City," Ergon Energy, Queensland. http:// www.townsvillesolarcity.com.au/Home/tabid/36/Default.aspx.

19 O. Gol, "Renewable Energy—Panacea for Climate Change?" Proceedings of ICREPQ'08, Santander, 12-14 March 2008.

20 Department of Resources, Energy and Tourism, Australian Government, "Clean Energy Initiative Overview," 2011. http://www.ret.gov.au/Department/Documents/cei/CEI_Fact_Sheet.pdf

21 V. C. Gungor, D. Sahin, T. Kocak, C. Buccella, C. Cecati and G. P. Hancke, "Smart Grid Technologies: Communication Technologies and Standards," IEEE Transactions on Industrial Informatics, Vol. 7, No. 4, 2011, pp. 529- 538.

22 V. C. Gungor, B. Liu and G. P. Hancke, "Opportunities and Challenges of Wireless Sensor Networks in Smart Grid," IEEE Transactions on Industrial Electron, Vol. 57, No. 10, 2010, pp. 3557-3564.

23 L. Wenpeng, D. Sharp and S. Lancashire, "Smart Grid Communication Network Capacity Planning for Power Utilities," Proceedings of IEEE PES Transmission Distribution Conference Exposium, New Orleans, 19-22 April 2010, pp. 1-4.

24 Q. Yang, J. A. Barria and T. C. Green, "Communication Infrastructures for Distributed Control of Power Distribution Networks," IEEE Transactions on Industrial Informatics, Vol. 7, No. 2, 2011, pp. 316-327.

25 V. C. Gungor, D. Sahin, T. Kocak and S. Ergut, "Smart Grid Communications and Networking," Turk Telecom, 2011.

26 IEEE Smart Grid, "Smart Grid Conceptual Framework." http://smartgrid.ieee.org/smart-grid-framework-diagram

27 C. W. Potter, A. Archambault and K. Westrick, "Building a Smarter Grid through Better Renewable Energy Information," Proceedings of IEEE/PES Power Systems Conference and Exposition, 15-18 March 2009, pp. 1-5.

28 "HOMER Energy." http://www.homerenergy.com/

29 "HOMER—Analysis of Micropower System Options." https:// analysis.nrel.gov/homer/

30 G. M. Shafiullah, A. T. Oo, D. Jarvis, S. Ali and P. Wolfs, "Prospects of Renewable Energy—A Feasibility Study in the Australian Context," Journal of Renewable Energy, Elsevier, Vol. 39, No. 1, March 2012.

31 I. H. Witten and E. Frank, "Data Mining: Practical Machine Learning Tool and Technique with Java Implementation," Morgan Kaufmann, San Francisco, 2000.

32 S. Ali and K. A. Smith, "On Learning Algorithm Selection for Classification," Journal on Applied Soft Computing, Vol. 6, No. 2, 2006, pp. 119-138.

33 R. Dimov, "Weka: Practical Machine Learning Tools and Techniques with Java Implementations," Technical Report, University of Saarland, Saarbrücken, 2007.

34 "Weka 3," Technical Report, The University of Waikato, New Zealand, 2009. http://www.cs.waikato.ac.nz/ml/weka/

35 G. M. Shafiullah, A. M. T. Oo, D. Jarvis, S. Ali and P. Wolfs, "Potential Challenges: Integrating Renewable Energy with the Smart Grid," Proceedings of Australasian Universities Power Engineering Conference, Christchurch, 5-8 December 2010, pp. 1-6.

36 "Renewable Energy Integration Facility," CSIRO. http://www.csiro.au/Outcomes/Energy/Renewable-Energy-Integration-Facility.aspx

37 G. M. Shafiullah, A. M. T. Oo, S. Ali, S. Azad, M. Arif and T. Moore, "Experimental Analysis of Harmonics on Utility Grid with PV Penetration," Proceedings of International Conference on Electrical and Computer Engineering (ICECE 2012), Ottawa, 22-24 August 2012, pp.

38 PSS Sincal, PSS Product Suite, Siemens. http://www.simtec.cc/ sites_en/sincal.asp

CHAPTER 7

Wave Tank Testing and Model Validation of an Autonomous Wave Energy Converter

Bret Bosma*, Tim Lewis, Ted Brekken and Annette von Jouanne

Wallace Energy Systems and Renewables Facility, Oregon State University, Corvallis, OR 97331, USA

ABSTRACT

A key component in bringing ocean wave energy converters from concept to commercialization is the building and testing of scaled prototypes to provide model validation. A one quarter scale prototype of an autonomous two body heaving point absorber was modeled, built, and tested for this work. Wave tank testing results are compared with two hydrodynamic and system models—implemented in both ANSYS AQWA and MATLAB/Simulink—and show model validation over certain regions of operation. This work will serve as a guide for future developers of wave energy converter devices, providing insight in taking their design from concept to prototype stage.

KEYWORDS

wave energy converter; physical modeling; numerical modeling

1. INTRODUCTION

In order to fully realize a robust, efficient, and cost-effective ocean wave energy converter, considerable modeling and testing of devices will be required. Due to the size and complexity of the full scale devices, the most cost effective way to make advances is through the use of numerical modeling and scaled prototype testing. This paper takes previous numerical modeling work and attempts to validate these models with a scaled prototype tested in a large wave flume.

Wave tank testing of wave energy converters is a complicated endeavor with many challenges. There is much to be learned from previous attempts at characterizing devices and validating models. The European Marine Energy Centre (EMEC) provides a tank testing standard in [1] and the University of Edinburgh has provided tank testing guidance in [2] and EquiMar has provided high level guidance in [3]. The book edited by Joao Cruz [4], has a chapter dedicated to numerical and experimental modeling of Wave Energy Converters (WECs) which helped to inform the process of setting up the experiment, and identify best practices.

The main thrust of this research is to outline the process of taking an idea of a WEC and bringing it through the prototype stage of development. This includes a significant amount of numerical modeling as well as physical modeling. The outcomes of this paper show the results of tank testing, namely the Response Amplitude Operators (RAOs) and power performance results compared with two different time-domain model approaches. This model validation helps to identify the regions of operation that can be reasonably modeled, allows for the adjustment of the model to more accurately match the real world, and provides confidence to change the model for future design iterations.

There is considerable interest in the development of ocean wave energy converters for remote sensing applications [5]. In fact, there has been interest in this for some time. Yoshio Masuda was a pioneer in the field, creating an oscillating water column navigation buoy, commercialized in 1965 [6]. Currently, atmospheric sensing equipment on board remote buoys is typically powered by a combination of solar, wind, and battery systems. Point absorber wave energy conversion technology would add to the sensing system power options. To this end, a heaving two-body point absorber was chosen as a possible solution, called an Autonomous Wave Energy Converter (AWEC). A target power level for full scale development was set at a nominal 200 W.

2. DEVICE GEOMETRY, SCALING, POWER TAKE OFF, AND MOORING

Many designs and geometry types exist for ocean wave energy extraction [6]. A heaving point absorber was chosen as the geometry to be modeled, built, and tested for this study. A point absorber is defined as a device whose horizontal extent is much smaller than one wavelength of an incoming wave. Energy conversion of point absorbers is limited by their heave only motion, however, benefits of this type of device include a relatively simple geometry, many previous studies and built devices based on this principle, and the robust nature of such a device.

2.1. Geometry

Full scale geometry design of the AWEC was outlined in [7], where coastal United States locations were chosen to help inform the design. A target of 200 W continuous power was chosen to meet the general electrical load of autonomous buoys. A focus on a simple shape was pursued for reasons of cost and ease of manufacture as well as ease of modeling.

A two-body approach was chosen where the relative motion between the bodies actuates the Power Take Off (PTO). Because of limitations related to the scaled testing facility, the need for additional stability within the system, and the shallow water depth of some prospective sites, a damping plate was added. To add further stability, the metacentric height was raised by the addition of weight to the bottom of the device, and buoyancy was added as high up as possible. This served to lower the center of gravity, while raising the center of buoyancy and thus creating a more stable device. The final-full scale equivalent, and actual fabricated geometry is shown in Figure 1, with the mass properties shown in Table 1.

2.2. Scaling

The prototype built was ¼ the full scale device design. To create the most representative model of the full scale device, with minimal modeling errors due to hydrodynamic and other nonlinear phenomena, the largest scale factor for the prototype was chosen. This was limited by what would reasonably fit in the testing facilities. A near shore location with water depth of about 14 m—a National Oceanic and Atmospheric

Administration buoy location off the coast of Galveston, TX—was used as a target location. The maximum water level in the proposed test facility divided by the water depth at the proposed site led to a scaling factor of four. That is to say that in terms of external physical dimensions, the target full scale device would be four times the physical size of the prototype. In scaling the model, ideally, two non-dimensional quantities, the Froude number and the Reynolds number would be held constant. However, for this test, viscous forces are assumed negligible and therefore the Froude scaling is assumed to be valid. Further investigation will inform more detailed viscous force models.

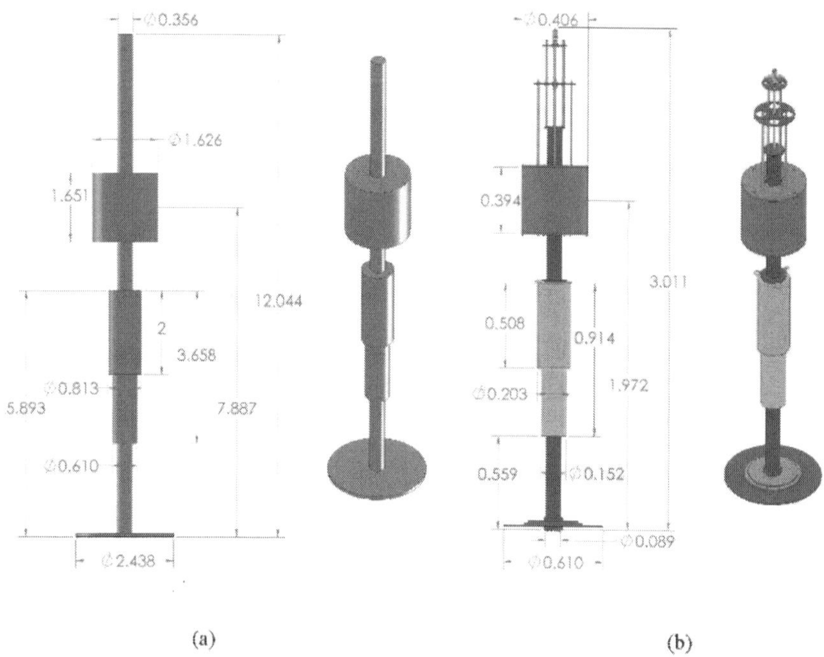

Figure 1. Autonomous Wave Energy Converter (AWEC): (**a**) full scale and (**b**) ¼ scale. All dimensions in meters. The float is toroidal shaped, half submerged in still water, around a central spar body.

Table 1. AWEC mass properties.

	Full scale		¼ Scale	
	Float	**Spar**	**Float**	**Spar**
Mass (kg)	1700	2250	26.5	35.2
Volume (m^3)	3.4	2.5	0.053	0.039

2.3. PTO

The PTO system is responsible for converting the force created by the motion of a WEC to some useful power. As is the case for a heaving point absorber, this often requires the translation of linear motion to rotary motion. There are many ways that this can occur including direct drive solutions as well as intermediate energy conversion solutions such as hydraulic systems. A study of direct drive solutions is performed in [8]. A linear generator system is detailed in [9,10]. For the project documented in this paper, a ball screw spindle drive was chosen coupled to a high efficiency brushed DC motor. At the scale of the built device, this provided the highest efficiency, lowest cost, and easiest implementation with readily available products. For example, one beneficial aspect of this PTO device was the readily available drive electronics consisting of a four quadrant servo-controller. This provided an easily adaptable platform for the implementation of various control schemes [9,10].

2.4. Mooring

The mooring of wave energy conversion devices is very important and should not be overlooked. Mooring can significantly affect the power production, survivability, environmental impact, and cost. An overview of a design approach is given in [11]. Mooring details for the scaled device are shown in Section 4.3. It is anticipated that for full scale open ocean deployment a single point catenary mooring would be favored.

3. LINEAR TEST BED TESTING

Due to the complex nature of a wave energy converter, there is a lot of useful information that can be gleaned from testing before the device ever enters the water. For this particular project the benefits of dry testing served two purposes. First, it allowed for the testing of most system components in a situation similar to those it sees in the water. Second, it allowed for the characterization of system losses, which are a significant contributing factor affecting the performance of a device. These losses can take many forms, but mostly can be attributed to friction, gearbox, and generator inefficiencies. In an attempt to characterize the system losses, as well as do a complete system validation, a Linear Test Bed (LTB) located in the Wallace Energy Systems and Renewables Facility was used as shown in Figure 2. An overview of the Linear Test Bed is given in [12].

Figure 2. Testing in the Linear Test Bed (LTB). The Linear Test Bed has a maximum stroke of 2 m, force capacity of 20 kN, and a maximum speed of 1 m/s.

3.1. Power Delivered to the Load

A useful measure of device performance is the electrical power delivered to the load for different operating conditions. A sweep of constant velocities and different load conditions were conducted as trials. Power was measured as the product of the voltage across and current through the DC motor load resistance. This was then scaled using Froude scaling to obtain the full scale equivalent power. DC motors have non-linear characteristics, such as static friction, windage losses, and hysteresis and eddy current losses. These can be included in complex models, but are second order effects and are not included here. Figure 3 shows the results from this test; velocity and power in the table are Froude scaled to full scale.

	0.54 m/s	0.68 m/s	0.90 m/s
3.5 ohm	222	344	619
7 ohm	269	415	748
10.5 ohm	267	415	749
14 ohm	251	394	706
28 ohm	188	294	529
56 ohm	118	186	332

Full Scale Equivalent Power (W)

Figure 3. Power delivered to the load in Watts. Power generally increases as velocity increases. A load resistance near the generator internal resistance (7.41 ohms) provides max power.

Note that the power generally increases as the velocity increases as expected. Also notice that the load resistance near the internal resistance of the generator (7.41 ohms) provides the max power from the generator. To illustrate this, considerFigure 4 which shows a basic dc motor circuit and power curve. From the maximum power transfer theorem the maximum power transfer occurs when the load resistance matches the internal generator resistance.

3.2. Loss Modeling

In order to estimate the total losses in the system, a series of tests were conducted. A profile was loaded into the LTB to cause a constant velocity between the float and the spar. For each trial, a velocity and a load condition were chosen, and the power losses were calculated as follows:

$$P_{load} = \left(\frac{e}{R_{DC} + R_{load}}\right)^2 R_{load} \quad (1)$$

$$P_{in} = P_k + P_r + P_{loss} + P_{elec} \quad (2)$$

$$P_{elec} - V_{gen} I_{gen} \quad (3)$$

where Pin is the power input to the system by the LTB; Pk is the rate of change of the linear kinetic energy present in the system which is zero for a constant velocity; Pr is the rate of change of the rotational kinetic energy in the system which is zero for a constant velocity; Ploss is the power losses in the system, and Pelec is the power measured out of the generator.

$$P_{loss} = P_{in} - P_{elec} \quad (4)$$

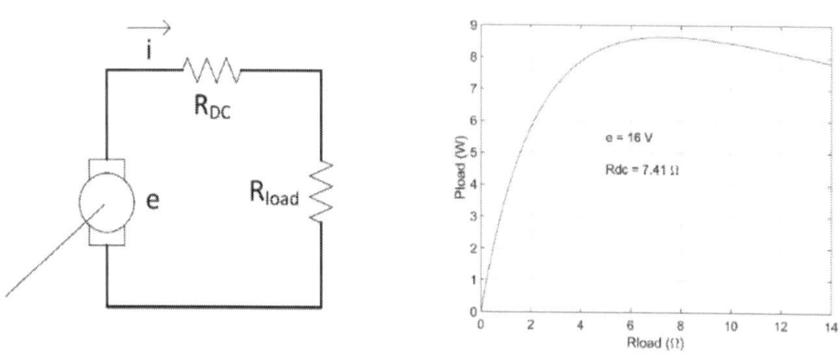

(a) (b)

Figure 4. (**a**) Basic DC motor model; (**b**) Theoretical maximum power transfer between motor and load. Maximum power occurs when the load resistance value is close to the internal motor resistance value.

The power input to the system Pin was measured as the force (measured with load cells) imparted on the device under test multiplied by the velocity.

$$P_{in} = F_{in}v \tag{5}$$

In order to include the losses in any model of the overall system, an estimate of those losses should be included. Although these losses are nonlinear, a linear damping term is a simple way to implement the estimated losses and is a good first pass approximation as shown in Chapter 13 in [13] and Section 2.3 in [14]. This term can be estimated as follows:

$$F_{loss} = B_{loss}v \tag{6}$$

$$P_{loss} = F_{loss}v = B_{loss}v^2 \tag{7}$$

$$B_{loss} = \frac{P_{loss}}{v^2} \tag{8}$$

where Floss is the equivalent loss force in the system; Bloss is the loss damping and captures friction and other high order loss mechanisms; v is the linear velocity; and Ploss is the power lost due to the inefficiencies of the system. Again, the velocity and load conditions were swept with the resulting estimates for Bloss shown in Figure 5; velocity and power in the table are Froude scaled to full scale. As expected, the higher speeds produced a lower estimated damping value. Based on these results, and anticipated tank testing velocities from 0.1 m/s to 0.7 m/s, a mean value of 2000 Ns/m was used in both AQWA and MATLAB/Simulink full scale models.

	0.54 m/s	0.68 m/s	0.90 m/s
3.5 ohm	2648	2262	941
7 ohm	2457	2219	889
10.5 ohm	2432	2032	770
14 ohm	2318	2039	824
28 ohm	2391	2275	757
56 ohm	2274	2068	819

Damping (Ns/m)

Figure 5. Estimated loss damping values for a sweep of velocity and load values. Higher speeds produced lower estimated damping values.

4. WAVE TANK TESTING FACILITIES

The next step in the model validation process was testing the scaled AWEC model in the large wave flume located in the O.H. Hinsdale Wave Research Laboratory (HWRL) at Oregon State University (OSU). The HWRL has performed similar wave energy converter testing in the large wave flume as detailed in [15], and shown in Figure 6, which made for a relatively smooth process for testing. Data from the test trials were recorded for the three major data acquisition systems included in the testing. This included the wave flume, optical tracking, and device performance data which will be reviewed in the following sections.

4.1. Large Wave Flume

The large wave flume is one of the largest of its kind in North America and measures 88 m long by 3.7 m wide and 4.6 m deep. The wave maker is hydraulic-piston actuated, and capable of creating regular, irregular, tsunami, and user-defined wave types. The period range is from 0.5 to 10 s with a maximum wave height of 1.6 m at a 7-second period. A 1:12 slope concrete beach was selected to minimize reflections within the tank. A water depth of 3.353 m was chosen as this was the maximum reasonable depth for testing in the flume.

For the AWEC testing setup, several data acquisition measurements related to the wave tank were included. A wavemaker start signal, wavemaker displacement, a wavegauge located at the wavemaker, and a pressure sensor level signal were all recorded. Additionally, three resistance-type self-calibrating wave gauges offset from the tank wall near the device under test were recorded. All signals were logged with a data acquisition system with a sampling rate of 50 Hz.Figure 6 shows the basic geometry and layout of the large wave flume, with the location of the wave gauges, device under test (model AWEC), and optical tracking system (PhaseSpace).

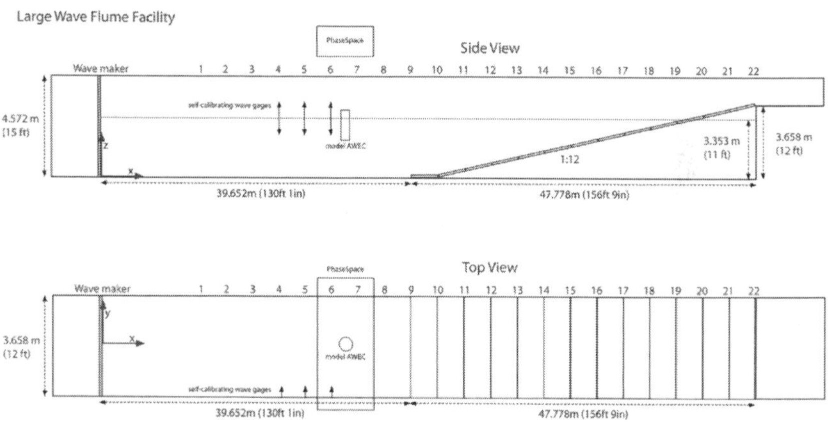

Figure 6. Large wave flume configuration.

4.2. Optical Motion Capture System

In attempting to validate numerical models of the AWEC system, it was very important to have confidence in the motion tracking data of the bodies under test. A PhaseSpace optical motion capture system was used for this purpose. This system, designed for the entertainment industry, allows high resolution tracking of rigid bodies. Twenty-seven uniquely identified active LED markers viewed by eight cameras surrounding the device describe the PhaseSpace setup. Motion data is recorded at a 480 Hz sampling rate. Previous tests found a 6-sigma accuracy of 0.9 mm for all targets within a 1.2 m radius, and 1.3 mm accuracy up to a 2.5 m radius [15]. Each body of the AWEC was uniquely identified for six degrees of freedom in relation to the reference coordinate definition square. Figure 7 shows the PhaseSpace setup on the left and a computer image showing the cameras and markers for each LED on the right.

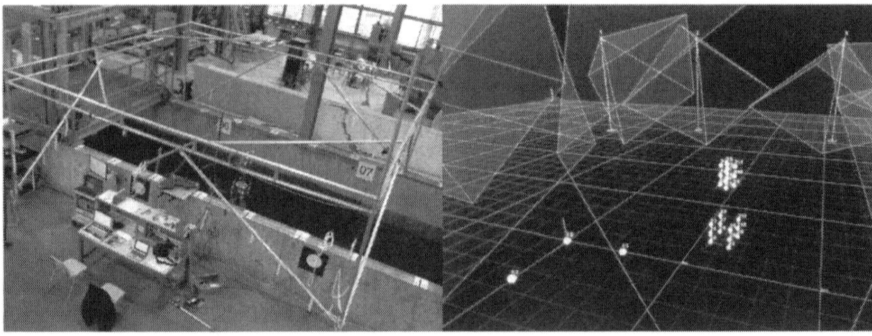

Figure 7. Optical motion tracking: (**a**) Setup in the O.H. Hinsdale Wave Research Laboratory (HWRL); and (**b**) screen capture of tracking markers.

4.3. Mooring

For testing in the large wave flume at the HWRL, a three point mooring system was constructed and implemented. Each line was installed in pretension in a horizontal plane with a 120° angle between the lines. The aft mooring line was parallel to the wave tank walls. Each line had a pulley system which led to a load cell on the edge of the tank. Mooring force was recorded for each line. Each line consisted of a 1.5 m un-stretched length of surgical tubing and a length of stiff rope. Experimental testing on the surgical tubing revealed an estimated stiffness of 520 N/m for a full scale equivalent, obtained using Froude scaling. Each line was set in a pretension of 260 N full scale equivalent (*i.e.*, pre-stretched 0.5 m). The surgical tubing was tested before connecting the mooring system and the stiffness was found to be suitably linear over the range of expected tensions. Assuming the *x*-*y* plane to be collinear with the three point mooring forces, *i.e.*, surge and sway, and the z-axis to be heave, the linearized stiffness about the equilibrium point is 780 N/m in *x* and *y* directions and 0 N/m in the *z*direction. Thus this mooring design will hold the model centered in the tank, yet not restrict the desired heave motion.

4.4. Device Performance Data

Device performance data from the AWEC was also recorded with the data acquisition system. In all, sixteen channels of data were recorded at a 50 Hz sampling rate using the HWRL data acquisition system. The signals can be broken into two categories, namely signals related to the PTO and signals related to the mooring.

Several signals related to the PTO, consisting of a lead screw, gearbox, generator, and encoder were logged. From the encoder data, position, velocity, and acceleration signals were recorded, which describe the relative motion between the two bodies: the float and spar. Voltages and currents were logged including the system bus voltage, power supply current, and load current. The general setup for the PTO circuit is shown in Figure 8. Several signals related to the 4-quadrant drive were logged including the drive current, drive velocity, commanded current, and the enable command. Additionally each of the three mooring lines had a load cell attached, and this mooring force was also recorded.

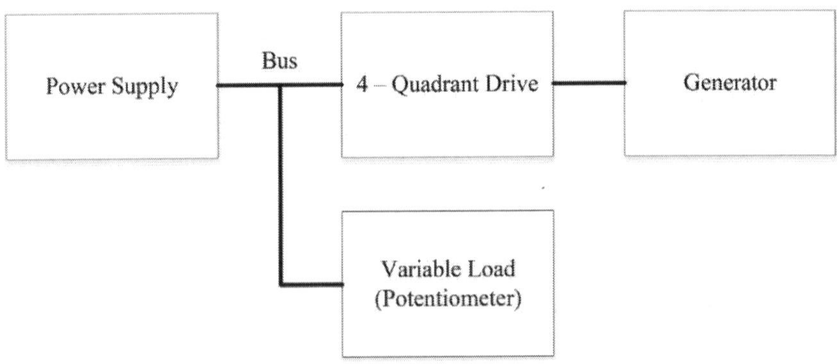

Figure 8. Power Take Off circuit.

5. WAVE TANK TESTING PROCEDURE

Wave tank testing occurred over a two week period in the HWRL. Three and a half days were dedicated to the setup of the wave tank. This included installation of the mooring system, installation of the device under test, installation of the optical motion tracking system, installation of the wave gauges and related wave tank systems, filling of the tank, and calibration of the system. Six days were then dedicated to experimental testing which is outlined below. Figure 9 shows the device under test.

Figure 9. AWEC testing in the Hinsdale Wave Research Laboratory at Oregon State University.

5.1. Experimental Methodology and Procedure

Each individual running of waves was considered to be a trial. Each trial was a member of an experiment. In all, there were nine experiments conducted with a total of one hundred and forty-three trials. All trials were run with either monochromatic or Pierson-Moskowitz spectral wave input. In general, with a few exceptions, the length of run was set to 200 waves (based on the peak period if spectral). In addition to the 200 waves, there was a ramp up time and ramp down time associated with the data acquisition and the wavemaker. Also in general, if control was to be implemented in the run, which was the case for most trials, the first ten percent of the trial was run with the control off. At ten percent of the total trial length the control was then activated and applied for the duration of the trial. This provided a baseline operation for each trial to see the effects of control on the system.

5.2. Types of Trials Conducted

Of the hundred and forty-three trials conducted throughout the testing, the bulk of the trials fit into four experiments. In an attempt to characterize the device, a sweep of wave height and period was conducted with a monochromatic input creating a five by five matrix. Also in characterizing the device, a sweep of irregular waves was performed, sweeping significant wave height and dominant wave period for a similar five by five matrix of values. The final experiments were conducted as an exercise in testing a binary and tertiary control scheme for generating more power from the AWEC. The results of these last two experiments will be published separately.

6. RESULTS

The main thrust of this work involved the model validation of a scaled prototype using numerical modeling techniques. Two numerical modeling tools, ANSYS AQWA and MATLAB/Simulink, were used to run full scale equivalent models of the AWEC. Results from the tank testing were scaled up for comparison with the numerical results using Froude scaling. Because of the discrepancies between the commanded and measured wave heights and periods, input to the numerical models were made to match full scale equivalent measured wave tank wave heights and periods.

6.1. Commanded vs. Measured Wave Tank Results

For regular wave inputs wave gauge readings were taken on the wave maker side of the AWEC. A desired period and wave height was given to the wave maker operator for each trial. The resulting wave surface elevation was recorded at the location of the wave gauge closest to the AWEC. All wave gauge data was filtered using a fourth order bandpass Butterworth filter with a passband between 0.1 Hz and 3 Hz. The resulting measurement of the wave period closely matched the expected period, however, the measured wave height deviated from the expected by as much as 29%. Furthermore, for different periods given the same expected wave height, the measured average wave height varied. Because of these results, all time domain simulations done in both AQWA and MATLAB/Simulink received the scaled measured input wave height in order to compare the theoretical results to measured results.

For irregular wave input, a similar process was followed. Because of restrictions on the operation of the wave maker, two of the trials in the matrix were unable to be completed. Results showed that there is significant error in both peak period and significant wave height with a max error of 27% and 35% respectively. In order to compare the hardware results with the numerical modeling, a scaled time series of measured wave heights was used as the input to the AQWA and MATLAB/Simulink numerical modeler.

6.2. Max Damping Value

For purposes of model validation and characterization of the system a fixed damping value was needed in order to conduct the tests. In order to achieve a fixed damping value the 4-quadrant drive was set in a current control (*i.e.*, torque control) operating mode. A velocity proportional damping term B (Ns/m) was then used to set the desired torque command to the motor. Early tests were run using a monochromatic wave input to the system. As the trial progressed, after approximately every ten waves the damping values were stepped up. Figure 10 shows the results from a damping sweep for two different monochromatic input waves, one with a full scale wave height of 1 m and period of 4 s, and one with a full scale wave height of 2 m and period of 6 s.

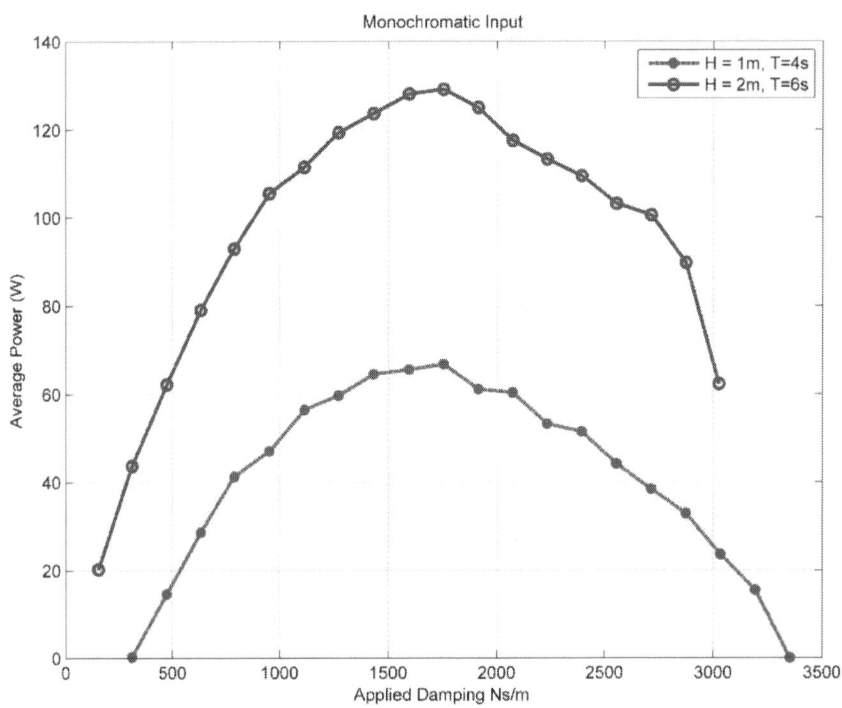

Figure 10. Average Power *vs.* Damping for two monochromatic wave inputs.

As the plot shows, the maximum full scale damping value for either case was near $B = 1600$ Ns/m, which was then chosen as the base damping value to be used for all sweeps of wave height and period, both monochromatic and spectral.

6.3. Response Amplitude Operators (RAOs)

Testing in regular waves gives insight into the behavior of the system in a relatively controlled environment, controlled by two parameters namely wave height and wave period. One measure of the performance of such regular wave tests is the calculation of the Response Amplitude Operator (RAO). The RAO is essentially a transfer function describing the relationship between an input and output characteristic of the components of the device of interest. Traditionally, in ship design, RAOs are often used in the design stage to determine modifications that may need to occur for safety reasons. In ocean wave energy conversion RAOs can be useful in design for maximum power extraction. For this study, because of the nature of the point absorber, the heave motion was identified as the pertinent motion for energy capture and thus the RAO was calculated for each body, namely the float heave and spar heave as follows:

$$RAO = \frac{\zeta}{\eta}\left[\frac{m}{m}\right] \tag{9}$$

where ζ is the body heave amplitude and η is the incoming wave amplitude. For each body the time series response was analyzed to find the positive zero crossings. Then for each period the amplitude of the waveform was calculated. The average of these values was then determined to be the body average heave amplitude. A similar process was used to obtain the average incoming wave amplitude.

As part of the model validation process, three sets of heave RAOs were calculated and compared for each rigid body. First, the tank test results of wave surface elevation as measured by the resistance wave gauges was used with the optical motion tracking information from PhaseSpace for the heave amplitude of each body. Second, time domain simulations from six degrees of freedom ANSYS AQWA-NAUT were used to calculate the heave RAO. Third, a heave-only time domain model was implemented in MATLAB/Simulink, based on the frequency domain parameters obtained from ANSYS AQWA LINE. The methodology and a review of this modeling process is outlined in [16]. The heave motion RAOs for the three data sets are shown in Figure 11. Results show the float acts close to a wave follower, that is to say, for a certain input wave height the response is close to the same in magnitude. Conversely, the response of the spar is greatly attenuated from the input waveform, causing relative motion between the two bodies, which is the mode for electricity generation. The AQWA and MATLAB results are in reasonable agreement and generally have the same shape with regard to frequency. However, both AQWA and MATLAB appear to be missing some high frequency behavior for the float and low frequency behavior for the spar present in the tank testing results. In the case of the float, wave steepness, which is higher at higher frequencies, may cause more nonlinear behavior and help explain the divergence in the results. Additionally, any unmodelled viscous forces, which are proportional to the square of speed, will be more significant. For the spar, the low frequency behavior could be accounted for by longitudinal resonant modes of the wave flume, which are not included in the AQWA and MATLAB models. The cause for the discrepancy will be investigated in future research.

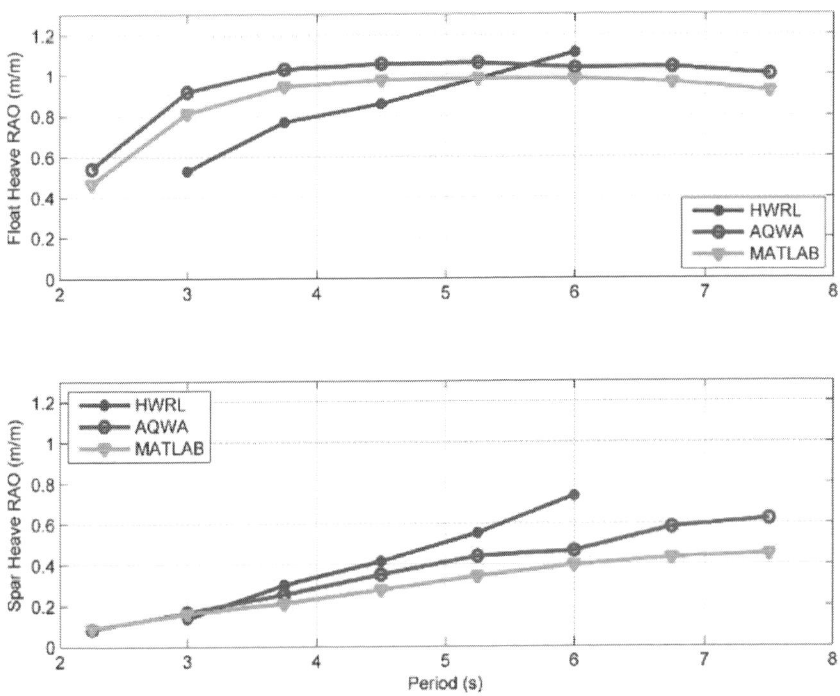

Figure 11. Response Amplitude Operators (RAOs) for HWRL, AQWA, and MATLAB. The PTO damping is set to 1600 Ns/m as described in Section 6.2.

6.4. Power Characteristics

One key measure of the performance of a wave energy converter is its average power output. This is particularly important for an autonomous application because a sensor package may rely on a certain baseline average power production for consumption. In characterizing a device it is most useful to look at the power output for irregular waves because this will be the case for any real world conditions. As outlined earlier, a sweep of significant wave height and peak period was done to create a 5 × 5 matrix of system outputs. Power production was calculated by measuring the bus voltage and the currents in the circuit and then calculating the power generated by the PTO. The result is the total power produced including all losses. Time domain simulations were then performed using ANSYS AQWA-NAUT and a time domain simulation tool in MATLAB/Simulink developed by the author, described in [17], in an attempt to simulate the testing environment. As described earlier, the losses implemented in the numerical models were estimated as additional damping in the system. The result is a power matrix plot showing the average power output for different combinations of significant wave height and peak period irregular wave inputs for one experimental test and two hydrodynamic modeling exercises as shown in Figure 12.

Hs(m)	Tp (s) 3.00	3.75	4.50	5.25	6.00
0.50	4.9	3.3	2.3	3.4	2.4
0.80	9.9	9.3	12.7	10.8	7.7
1.10	31.6	41.3	32.3	24.6	20.2
1.40		66.7	58.4	44.7	33.2
1.70		83.6	99.6	66.2	52.3
	HWRL Pavg (W)				

(a)

Hs(m)	Tp (s) 3.00	3.75	4.50	5.25	6.00
0.5	7.9	7.9	6.2	5.1	4.4
0.8	21.9	21.2	17.0	14.8	11.7
1.1	34.2	37.7	32.3	28.0	23.2
1.4		56.8	49.3	42.7	36.8
1.7		72.2	72.3	64.2	52.6
	AQWA Pavg (W)				

(b)

Hs(m)	Tp (s) 3.00	3.75	4.50	5.25	6.00
0.5	11.4	8.8	6.6	5.0	4.3
0.8	32.5	24.9	16.9	14.0	10.4
1.1	44.7	41.5	32.3	25.5	22.9
1.4		62.2	53.0	42.0	35.8
1.7		70.4	72.8	64.5	49.5
	MATLAB Pavg (W)				

(c)

Figure 12. Power Matrix for irregular wave input: (**a**) HWRL wave tank testing; (**b**) AQWA model; and (**c**) MATLAB model. All quantities are Froude scaled to full scale.

In order to quantify the differences between the model and the measured data, the percent error was calculated between the different methods. Figure 13 shows the % error between the measured data and AQWA simulated data. As shown, the match is quite good at the nominal case. As lower significant wave heights and peak periods are analyzed the error generally gets larger, with the AQWA model over-predicting the power output of the system. As significant wave heights increase and peak periods decrease the trend shows an AQWA under-prediction of power output. This may be due to nonlinear PTO characteristics such as friction which may be more dominant at the lower significant wave heights and lower WEC speeds. These effects were not included in the numerical model as a constant linear damping term was used.

Hs(m)		Tp (s)				
		3.00	3.75	4.50	5.25	6.00
	0.50	60.8	138.9	167.3	49.3	85.2
	0.80	122.1	128.1	34.0	37.2	52.5
	1.10	8.3	-8.8	0.0	13.9	14.6
	1.40		-14.8	-15.5	-4.5	10.6
	1.70		-13.6	-27.4	-2.9	0.6
		% error between HWRL and AQWA				

Figure 13. Percent error between measured tank testing data and AQWA model data. AQWA over-predicts for lower significant wave heights and peak periods. AQWA under-predicts for higher significant wave heights and peak periods. All quantities are Froude scaled to full scale.

7. CONCLUSIONS

This work provides an overview of linear test bed and wave tank testing procedures, as well as model validations stemming from the tank testing results. Design and testing methodologies were overviewed and results for characterizing a WEC and validating a model were given. Results include power output results from irregular wave input, and response amplitude operator (RAO) results from regular wave input. Numerical methods were shown to have relatively accurate results around a nominal operating point with increased error in the model as the conditions varied from the nominal. Tank testing of the AWEC showed under sufficiently energetic wave conditions electrical power generation performance met the target. Future work includes refinements to the hydrodynamic design and the PTO system to improve power generation under more nominal conditions.

AUTHOR CONTRIBUTIONS

B.B., T.L., A.J. and T.B. conceived and designed the experiments; B.B. and T.L. performed the experiments; B.B., T.L., A.J. and T.B. analyzed the data; B.B. wrote the paper.

CONFLICTS OF INTEREST

The authors declare no conflict of interest.

REFERENCES

1 Holmes, B. Tank Testing of Wave Energy Conversion Systems. European Marine Energy Centre (EMEC): Orkney, UK. Available online: http://www.emec.org.uk/tank-testing-of-wave-energy-conversion-systems/ (accessed on 3 December 2012).

2 Payne, G.; Ingram, D. Best practice guidelines for tank testing of wave energy converters. *J. Ocean Technol.* **2009**, *4*, 38–70.

3 High Level EquiMar Protocols—EquiMar. Available online: http://www.equimar.org/high-level-equimar-protocols-.html (accessed on 1 November 2013).

4 Cruz, J. *Ocean Wave Energy: Current Status and Future Prespectives*, 1st ed.; Springer: Berlin, Germany, 2008.

5 Hart, P. Autonomous Powerbuoys: Wave Energy Converters as Power Sources for the Next Generation of Ocean Observatories. Available online: http://virtual.ocean-news.com/article/Autonomous_Powerbuoys%3A_Wave_Energy_Converters_As_Power_Sources_For_The_Next_Generation_Of_Ocean_Observatories/1081385/114319/article.html(accessed on 9 May 2013).

6 Falcão, A.F.O. Wave energy utilization: A review of the technologies. *Renew. Sustain. Energy Rev.* **2010**, *14*, 899–918.

7 Lewis, T.M.; von Jouanne, A.; Brekken, T.K.A. Modeling and control of a slack-moored two-body wave energy converter with finite element analysis. In Proceedings of the 2012 IEEE Energy Conversion Congress and Exposition (ECCE), Raleigh, NC, USA, 15–20 September 2012; pp. 938–945.

8 Rhinefrank, K.; Schacher, A.; Prudell, J.; Brekken, T.K.A.; Stillinger, C.; Yen, J.Z.; Ernst, S.G.; von Jouanne, A.; Amon, E.; Paasch, R.; *et al*. Comparison of Direct-Drive Power Takeoff Systems for Ocean Wave Energy Applications. *IEEE J. Ocean. Eng.* **2012**, *37*, 35–44.

9 Bostrom, C.; Waters, R.; Lejerskog, E.; Svensson, O.; Stalberg, M.; Stromstedt, E.; Leijon, M. Study of a Wave Energy Converter Connected to a Nonlinear Load. *IEEE J. Ocean. Eng.* **2009**, *34*, 123–127.

10 Bostrom, C.; Ekergard, B.; Waters, R.; Eriksson, M.; Leijon, M. Linear Generator Connected to a Resonance-Rectifier Circuit. *IEEE J. Ocean. Eng.* **2013**, *38*, 255–262.

11 Johanning, L.; Smith, G.H.; Wolfram, J. Mooring design approach for wave energy converters. *J. Eng. Marit. Environ.* **2006**, *220*, 159–174.

12 Hogan, P.M. A Linear Test Bed for Characterizing the Performance of Ocean Wave Energy Converters. Master's Thesis, Oregon State University, Corvallis, OR, USA, 2007.

13 Mohan, N.; Robbins, W.P.; Undeland, T. *Power Electronics*; John Wiley & Sons: Hoboken, NJ, USA, 1995.

14 Krause, P.C.; Wasynczuk, O.; Sudhoff, S.D.; Pekarek, S. *Analysis of Electric Machinery and Drive Systems*; John Wiley & Sons: Hoboken, NJ, USA, 2013.

15 Rhinefrank, K.; Schacher, A.; Prudell, J.; Hammagren, E.; Stillinger, C.; Naviaux, D.; Brekken, T.; von Jouanne, A. Scaled wave energy device performance evaluation through high resolution wave tank testing. In Proceedings of the OCEANS 2010, Seattle, WA, USA, 20–23 September 2010; pp. 1–6.

16 Bosma, B.; Zhang, Z.; Brekken, T.K.A.; Ozkan-Haller, H.T.; McNatt, C.; Yim, S.C. Wave energy converter modeling in the frequency domain: A design guide. In Proceedings of the 2012 IEEE Energy Conversion Congress and Exposition (ECCE), Raleigh, NC, USA, 15–20 September 2012; pp. 2099–2106.

17 Bosma, B.; Brekken, T.; Ozkan-Haller, H.T.; Yim, S.C. Wave Energy Converter Modeling in the Time Domain: A Design Guide. In Proceedings of the IEEE Conference on Technologies for Sustainability, Portland, OR, USA, 1–2 August 2013; pp. 103–108.

CHAPTER 8

Fundamental Open Questions on Engineering of "Super" Hydrogen Sorption in Graphite Nanofibers: Relevance for Clean Energy Applications

Yury S. Nechaev[1*], Alp Yürüm[2], Adem Tekin[3], Nilgün Karatepe Yavuz[4], Yuda Yürüm[5], T. Nejat Veziroglu[6]

[1]Bardin Institute for Ferrous Metallurgy, Moscow, Russia

[2]Nanotechnology Research and Application Center, Sabanci University, Istanbul, Turkey

[3]Informatics Institute, Istanbul Technical University, Istanbul, Turkey

[4]Energy Institute, Istanbul Technical University, Istanbul, Turkey

[5]Faculty of Engineering and Natural Sciences, Sabanci University, Istanbul, Turkey

[6]International Association for Hydrogen Energy, Miami, USA

ABSTRACT

Herein, some fundamental open questions on engineering of "super" hydrogen sorption (storage) in carbonaceous nanomaterials are considered, namely: 1) on thermodynamic stability and related characteristics of some hydrogenated graphene layers nanostructures: relevance to the hydrogen storage problem; 2) determination of thermodynamic characteristics of graphene hydrides; 3) a treatment and interpretation of some recent STM, STS, HREELS/LEED, PES, ARPS and Raman spectroscopy data on hydrogensorbtion with epitaxial graphenes; 4) on the physics of intercalation of hydrogen into surface graphene-like nanoblisters in pyrolytic graphite and epitaxial graphenes; 5) on the physics of the elastic and plastic deformation of graphene walls in hydrogenated graphite nanofibers; 6) on the physics of engineering of "super" hydrogen sorption (storage) in carbonaceous nanomaterials, in the light of analysis of the Rodriguez-Baker extraordinary data and some others. These fundamental open questions may be solved within several years.

KEYWORDS

Hydrogen "Super" Sorption (Storage), Graphenes, Carbon Nanotubes, Graphite Nanofibers, A Breakthrough Hydrogen Storage Nanotecnology, Clean Energy Applications

1. INTRODUCTION

Hydrogen is currently one of the most promising "green" fuels, owing to the fact that its gravimetric (mass) energy density of 142 MJ/kg (39.5 kWh/kg), (Figure 1, [1]) exceeds that of petroleum (oil) by a factor of three and that the product of its combustion is water vapor. On the other hand, the volumetric (volume) energy density of molecular gaseous hydrogen at 1 bar pressure is lower by several orders than that of oil (Figure 1, [1]), but it can exceed the oil quantity at megabar pressures (Figure 1, [1] - [3]). In light of this, the issue of finding systems and materials for a compact and energy efficient hydrogen storage assumes a primary importance. The known data on volumetric (volume) and gravimetric (mass) energy densities for different energy carriers [1] are shown in Figure 1. The additional symbol (the red circle) corresponds to the analytical data [2] [3] on the molecular solid hydrogen intercalated into the hydrogenated graphite nanofibers (GNFs); those are related to the extraordinary data [4] - [12] (being under consideration in this paper). As is noted in a number of studies, for instance in the recent review [13] [14] , hydrogen-based fuel cells are promising solutions for the efficient and clean delivery of electricity. Since hydrogen is an energy carrier, a key step for the development of a reliable hydrogen-based technology requires solving the issue of efficient storage and transport of hydrogen. During the last few decades several proposals based on the design of advanced materials such as metal hydrides and carbon structures have been made to overcome the limitations of the conventional solution of compressing or liquefying of hydrogen in tanks.

Nevertheless, none of the proposed systems, with the exception of the nobody reproduced extraordinary experimental data [4] - [12] and the related analytical data [2] [3] , are currently offering the required performances for the on-board hydrogen storage in fuel-cell-powered electrical vehicles [15] [16] . The performances are usually formulated in terms of hydrogen storage gravimetric (mass) and volumetric (volume) capacities (Figure 2, [17]) and control of adsorption/desorption processes, particularly, relevance to so called "reversibility" of the stored hydrogen. Therefore the problem of hydrogen efficient storage remains so far unsolved and it continues to represent a significant bottleneck to the advancement and proliferation of fuel cell and hydrogen technologies [1] [13] - [19] .

It is shown [17] , with regards to the known achievements, that the US Department of Energy (DOE) targets for the on-board hydrogen storage systems in fuel-cell-powered vehicles [15] , relevant to gravimetric (mass) and volumetric (volume) storage densities (capacities) of "reversible" hydrogen for 2010 (6.0 mass% H_2(of the system mass), 45 kg $(H_2)/m^3$ (system)) and for 2015 (9.0 mass% H_2 (of the system mass), 81 kg $(H_2)/m^3$ (system)). These values are higher than the targets established by the E.U and/or Japan. The additional symbol (the large red circle) corresponds to the analytical data [2] [3] on the molecular solid hydrogen intercalated into the hydrogenated graphite nanofibers (GNFs); those are related to the extraordinary data [4] - [12] (being under consideration in this paper). The 2017 targets [16] are of 5.5 mass% H_2 (of the system mass) and 40 kg $(H_2)/m^3$ (system). As has been noted in [18] [19] : 1) "... to realize a compact and energy efficient hydrogen storage is a key technology" [18] ; 2) "... breakthroughs in hydrogen densities are strongly required" [18] ; 3) it is necessary "to find a breakthrough technology" [19] .

2. SOME OF THE MATERIALS UNDER STUDY IN COE'S

There are shown in Figure 3 the prognosis data [18] (amended in January 2011) on market entry of fuel cell powered vehicles (FCVs) and hydrogen charging stations. Particularly, it is shown that year 2015 can be related to the target commercialization start of FCVs to general public (see further shown Items 6 and 7—the Toyota announcement—2014), and Year 2025—to an increase numbers of FCVs and hydrogen stations based on profitable business. As is also shown (Figure 3, [18]), the present time is a very suitable (in the plan of the market entry prognosis) for such a developments. In this connection, it's also expedient to note about a number of communications of 2013-2014 (in Internet) on the nowadays market situation, relevance to fuel-cell electrical vehicles (FCVs) and hydrogen charging stations, for instance the following.

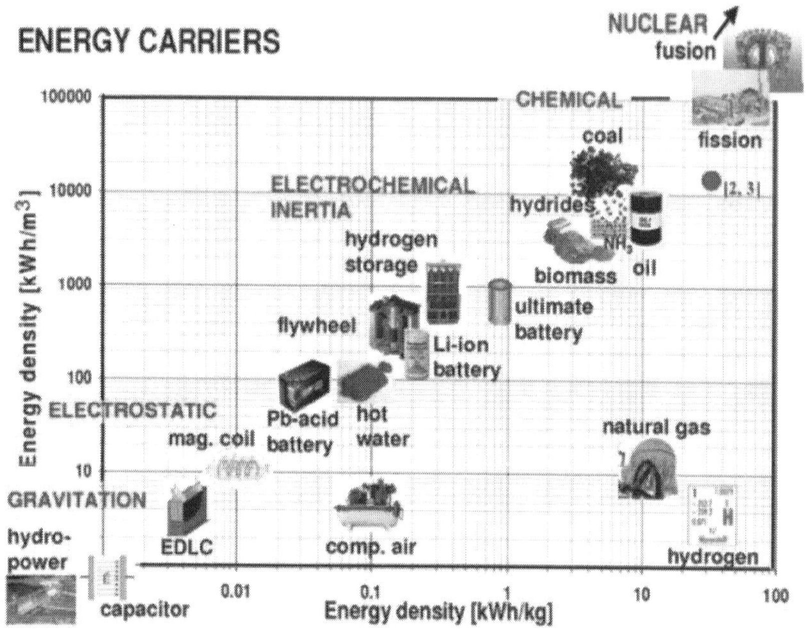

Figure 1. Data on volumetric (volume) and gravimetric (mass) energy densities for different energy carriers.

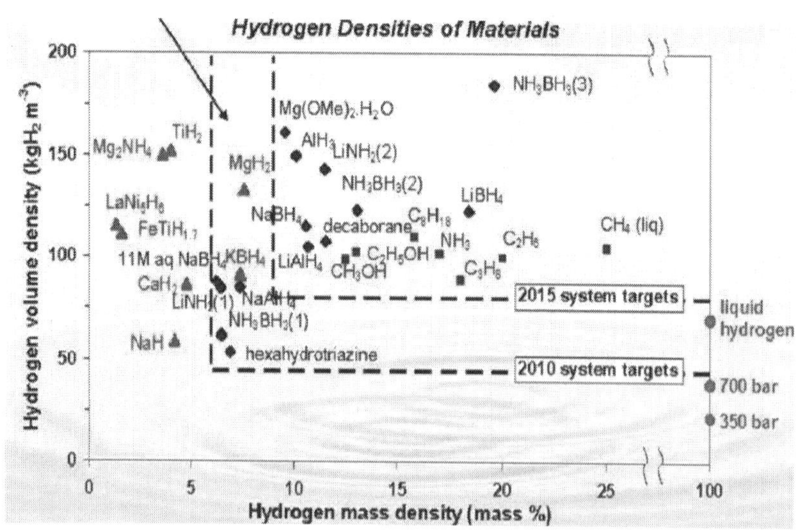

Figure 2. Correlation between volume and mass of hydrogen densities of various materials.

2.1. Hyundai's fuel-cell vehicle Could Be a massive success. The Motley Fool

Oct. 24, 2013: Hyundai isn't the only manufacturer in the race for FCVs. Toyota Motors, Daimler's Mercedes-Benz, BMW, and a number of other car companies have spent billions in fuel-cell technology, and are all competing to see who can be the first to market with a consumer-friendly FCV.

2.2. Hydrogen and Fuel Cells: GM—Honda Collaboration on Next-Generation Fuel Cell Technologies (03.07.2013)

(http://www.netinform.net/h2/Aktuelles_Detail.aspx?ID=3285.) Goal is commercially feasible fuel cell and hydrogen storage in 2020 time frame.

On July 2, 2013: General Motors (NYSE: GM) and Honda (NYSE: HMC) announced a long-term, definitive master agreement to co-develop next-generation fuel cell system and hydrogen storage technologies, aiming for the 2020 time frame. The collaboration expects to succeed by sharing expertise, economies of scale and common sourcing strategies. Source: www.gm.com.

2.3. Ballard signs long-term engineering services contract to advance Volkswagen AG Fuel Cell Automotive Research Program Ballard Power

Mar. 7, 2013: Ballard Power Systems has announced signing of an agreement with Volkswagen Group for a major Engineering Services contract to advance development of fuel cells for use in powering demonstration cars in Volkswagen's fuel cell automotive research program. The contract term is for 4 years, with an option for an extension. The expected contract value is in the range of $60 - $100 million.

2.4. Hydrogen fuel research May benefit from shift in auto industry (Hydrogen Fuel News)

Feb. 7, 2013: Ford, Daimler, and Nissan recently joined forces to make progress in the field of hydrogen fuel cell technology. Each of the companies has an ambitious goal in mind in terms of hydrogen transportation, but each has also been faces with challenges that threaten to derail these goals. The automakers decided to team up in order to overcome some of these challenges and introduce new standards to hydrogen fuel cell technology as a whole.

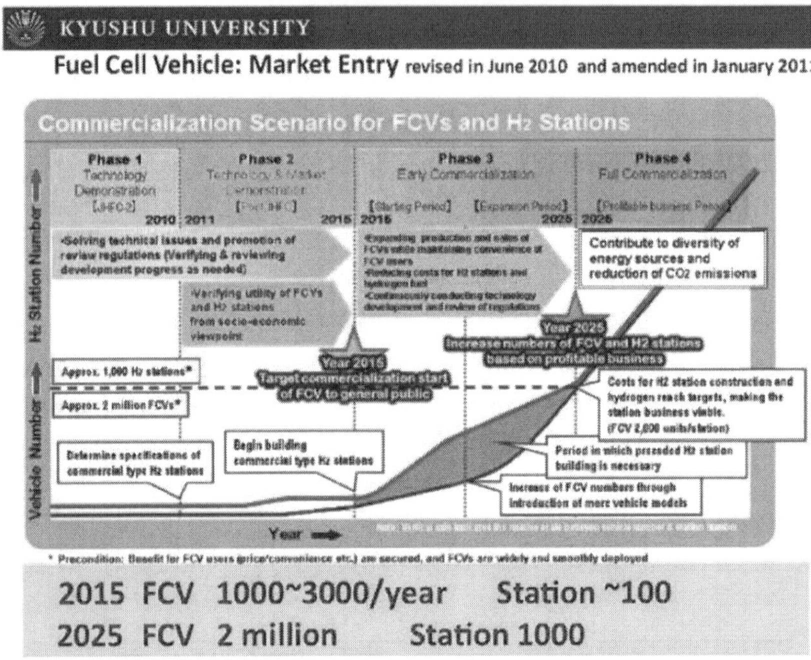

Figure 3. The prognosis data on market entry of fuel cell powered vehicles and hydrogen charging stations.

2.5. US Department of Energy launches new hydrogen fuel initiative (Hydrogen Fuel News)

May 23, 2013: Hydrogen fuel has become a major focus for the global auto industry and this focus is likely to transform the transportation sector around the world. As automakers put more emphasis on clean transportation, global markets are beginning to respond by supporting the promotion of hydrogen fuel in the transportation sector. Much of this support comes in the form of governments working to establish a working hydrogen fuel infrastructure that will be capable of supporting a new generation of fuel cell vehicles.

2.6. As Is noted in "CHFCA Weekly Fuel" of January 2014, Toyota unveils zero-emissions hydrogen fuel-cell "Car of the Future" for sale next year (Think Progress)

Toyota announced the launch of a hydrogen-powered fuel-cell car in the US next year at the annual Consumer Electronics Show (CES) in Las Vegas. The car, which resembles the popular Corolla, is yet to be named, but like the birth of a royal child it's the pedigree that counts—and Toyota is the largest auto manufacturer in the world.

2.7. As Is noted in "CHFCA Weekly Fuel" of January 2014, Toyota touts hydrogen fuel vehicles despite criticism (The Detroit News)

A top US Toyota Motor Corp. executive strongly stood by its focus on hydrogen fuel cell vehicles, defending their safety and dismissing criticism from the top executives at Tesla Motors, Nissan Motor Co. and others. "I realize there is no shortage of naysayer regarding the viability of this technology and the infrastructure to support it," said Bob Carter, senior vice president for automotive operations, Toyota Motor Sales USA Inc.

In connection with this, it is also expedient to take into account the long-term corporation (partnership) of Shell and General Motors Companies, relevance to hydrogen charging stations and fuel cell electrical vehicles, for instance:

1) July 2011: "The hydrogen infrastructure for automobiles is economically viable and do-able," said Larry Burns, Ph.D., General Motors Vice President, Research & Development and Strategic Planning… Shell Hydrogen has been developing hydrogen and fuel cell businesses since 1999 www.minichamps.ru/2011/general-motors-shell-fuel-up-on-hydrogen-in-los/angeles.

2) It was installed in November 2007 by Shell and General Motors to provide a venue for demonstrations to federal lawmakers and officials. Hydrogen fuel cells have long been touted as the next great energy revolution: www.thelivingmoon.com/41pegasus/02files/Alternate_Fuel_Shell_Oil_Hydrogen.html.

3) November 2004: The hydrogen-dispensing pump is the first installed at a public gas station in the country, according to officials from Shell Hydrogen

and General Motors Corp., who will team up today to… Virtually all the major auto manufacturers have prototype vehicles that run on fuel cells and are refining the technology: www.washingtonpost.com/wp-dyn/articles/A38168-2004Nov9.html.

4) Washington DC, March 5, 2003—General Motors Corp. and Shell Hydrogen are combining resources to help make hydrogen fuel cell vehicles a commercially viable reality, the two announced today. "The partnership brings together two leaders in hydrogen energy and transportation to take a coordinated, comprehensive approach," said Donald Huberts, chief executive officer of Shell Hydrogen. "By combining GM's expertise in vehicle technology with Shell's leadership in refueling technologies, the initiative represents an important step forward in the commercialization of hydrogen fuel cell vehicles" http://www.wec.org/news/shell-gm-partner-to-make-hydrogen-fuel-cell-vehicles-a-reality.

2.8. Shell predicts the end of the gas guzzler by 2070. The New Zealand Herald (2014)

Shell has released a report predicting the end of petrol-powered cars will be in 2070. The oil giant have compiled a 46-page report, using the progress in automotive fuel technology and economic scenarios as a basis for their prediction of all petrol cars becoming a thing of the past by 2070.

In the present paper are considered, in the light of a constructive critical analysis [2] [3] of a number of data [4] - [12] [20] - [22] , some fundamental aspects—open questions, relevant to developing, may be, within several years time frame, of a key breakthrough technology of a compact and energy efficient hydrogen storage in hydrogenated graphite nanofibers (GNFs).

2.9. The Northeastern University (NU) Group [4] - [8] Hydrogen Storage Activity

As was noted in [20] (2002) by Maeland, whose two works were cited in [4] (Ref. [1] (1978) and Ref. [4] (1981) in it), at the 1996 fall meeting of the US Materials Research Society (MRS), held in Boston, Massachusetts, Rodriguez and Baker of Northeastern University (NU), Boston, presented a paper in which they claimed the development of a "super" hydrogen storage nanomaterial. The material, graphite nanofibers (GNFs), discovered by Baker back in 1972 was claimed to be capable of storing up to 30 liters of gaseous hydrogen (H_2) per gram of nanofibers of the herringbone-like structure (Ref. [50] (1996) and Ref. [51] (1997) in [20]), i.e. about 73 mass% (of the graphite-hydrogen system mass) corresponding to formula CH32. The graphite nanofibers (GNFs) were preparing by reacting hydrocarbons with carbon monoxide on catalytic particles of bi- or tri-metallic nickel or iron. Hydrogen uptakes were determined by exposing, in a system of known volume, a purified GNF bundle (batch-like) sample to molecular gaseous hydrogen at room temperature, and observing the drop in pressure over a 24 hour period from an initial value of 11.2 MPa [4] (Ref. [61] (1998) in [20]). Three of the herringbone-like structure samples, according to [4] , take up hydrogen to give the hydrogen adsorption data of 62 ± 5 mass% (but not 73 mass%, as it was declared in the earlier talks (1996, 1997)), two of the platelet-like structure samples take up hydrogen to give the hydrogen adsorption data of 46 and 54 mass%, and the tubular-like structure sample— 11 mass%. Four of the herringbone-like structure hydrogenated samples, according to [4] , released the most of the stored hydrogen (at room temperature for 5 - 10 min) to give the reversible hydrogen desorption data of 48 ± 5 mass%; the rest part of the stored hydrogen was released at higher temperatures (under temperature-programmed desorption examination) to give the irreversible hydrogen desorption data of 14 ± 5 mass%. The results reported by Rodriguez and Baker immediately caused controversy. Michal Heben of the National Renewable Laboratory in Denver, Colorado, USA, pointed out that the highest ratio of hydrogen to carbon found in Nature is 4/1 (CH_4) and corresponds to 25 mass% and expressed skepticism (Ref. [51] (1997) in [20]) of the results and attempts to verify the results of Rodriguez and Baker. Ahn et al. (Ref. [52] (1998) in [20]) measured hydrogen adsorption and desorption at 300 K on graphite nanofibers (GNFs) and reported that the absolute level of hydrogen desorption from these materials were typically less than 0.025 H/C (0.2 mass%) which is comparable to other forms of carbon. Jarvi et al. (Ref. [53] (1999) in [20]) reported very low hydrogen storage capacity at 303 K, comparable to activated carbon, for graphite nanofibers (GNFs) prepared by catalytic decomposition of ethylene over nickel, iron, copper/nickel and alumina/magnesia catalysts and concluded that they were unlikely storage materials for hydrogen. However, they left the door open by stating that subtle processing effect might convert inactive materials into effective hydrogen sorbents (as has been recently shown in analytical studies [2] [3] , and is considered in the present analytical study). As was also noted in [20] , the announcement at the MRS meeting did not escape the automakers and Daimler-Chrysler began an evaluation study with the NU group of these "super" hydrogen storage materials. Later, however, Daimler-Chrysler ended their participation in the study. Then, Ford Motor Company was supporting the NU group (Ref. [54] (1999) in [20]). The NU group was also supported by the DOE in USA (Ref. [55] (1998) in [20]), but the support was terminated presumably because of the un-willingness of Rodriguez and Baker to share their GNF samples with other DOE laboratories for examination. More detailed description of the situation was done by Jennifer Babson, a freelance writer in Boston, who interviewed (in

November 1997) Rodriguez and Baker; see two 10/25/97 articles from the Economist on hydrogen fuel. This un-willingness to submit samples to other investigations had continued to fuel the controversy and prompted Dr. Gary Sandrock, a well known expert in the field of hydrogen storage materials, to publically call on Rodriguez to submit samples to others for a "Real-Word Test" of her nanofiber materials (Ref. [56] (1998) in [20]). It, however, had not been done (so far as we know) up to 2005 [21] [22] , despite the fact that Rodriguez and Baker were issued two related US patents of 1997 ("Storage of hydrogen in layered nanostructures") [6] and 2000 ("Method for introducing hydrogen into layered nanostructures") [7] and had thus secured protection for their process. As can be shown, the negative test-results [21] [22] , with respect to [4] -[12] data (i.e., both for Rodriguez-Baker et al. data [4] -[8] , and for Gupta et al. ones [9] -[12]), could be caused by using in [21] [22] the non-adequate GNF samples (including samples supplied for this test by Rodriguez and Baker themselves). The work of the NU group had been presented in a number of talks (Refs. [58]-[60] (1999) in [20]) and in two yearly cited (up to nowadays) articles in the Journal of Physical Chemistry of 1998 [4] and 1999 [5] . Article [4] has been cited (from 1998) 168 times in Scopus; the most recent citing is in two articles of 2013 [23] [24] . Article [5] has been cited (from 1999) 196 times in Scopus; the most recent citing is in two articles of 2013 [25] [26] .

Unlike [4] [5] (1998, 1999), the Gupta et al. papers [9] -[11] [12] (2000-2004, 2006) have not been discussed and/or cited so much, despite of the situation that, as far as we know, only experimental results [9] -[12] confirm (and reproduce, in an essential degree) the extraordinary experimental data [4] -[8] . And as far as we know, both authors [4] -[8] (1998-2000, 2005) and authors [9] -[12] (2000-2004, 2006) have never crossed out their extraordinary experimental results. Nevertheless, authors [4] (1998) had done some corrections (and/or modifications) in [5] (1999) of their original adsorption-desorption data. Baker (in [8] (2005)) had modified the reversible hydrogen adsorption-desorption data [4] (for the herringbone-like structure GNF samples) up to value of "40% by mass of molecular hydrogen per gram of carbon" that corresponds to 29 mass% of hydrogen (of carbon-hydro- gen system mass), instead of 48 ± 5 mass% quantity declared in [4] . In article of Lueking et al. [27] (2004), where Rodriguez and Baker were co-authors, it was noted the quantity of 67 mass%, relevance to [4] [6] data (for the herringbone-like structure GNF samples), and the quantity up to 40%, relevance to [5] [7] data (for the herringbone-like structure GNF samples). The graphite nanofibers (GNFs) possessing a herringbone-like structure and a high degree of defects (dislocations) were found [27] to exhibit the best performance for hydrogen storage resulted in 3.8 mass% release after exposure at 69 bar and room temperature (for 10 h). This result is in contrary with data [4] -[8] . But, as was stressed in [27] , the "herringbone" graphite nanofibers (GNFs) used in that investigation were produced from a different catalyst formulation than that used in [4] and [5] . Furthermore, as was also stressed in [27] , the hydrogenation adsorption/desorption protocol followed there was not the same as that used in [4] and [5] . It is necessary to emphasize that the extraordinary experimental results [4] -[12] have not been reproduced by other research teams worldwide [20] -[22] [27] -[30] . But the rather known authors [4] -[12] (see, for instance, information about them in Scopus and/or ScienceDirect.com Internet programs) have not definitely crossed out their data (as far as we know). Some of these works [4] [5] have been yearly cited (up to nowadays). On the other hand, the physics of such results [4] -[12] has been developed (in an essential degree) in analytical studies [2] [3] [31] -[33] . Therefore, there are serious reasons to assume (following to [34] [35]) that experimental works [4] -[12] have contained some "know-how" methodological and/or technological elements. It is under a consideration in the present analytical study.

3. SOME FUNDAMENTAL OPEN QUESTIONS OF THE CONSIDERED PROBLEM

3.1. On Thermodynamic Stability and Related Characteristics of Some Hydrogenated Graphene Layers Nanostructures: Relevance to the Hydrogen Storage Problem

The analytical study [2] [3] is devoted to consideration of the thermodynamic stability, and related thermodynamic characteristics of the following graphene layers systems: 1) double-side hydrogenated graphene of composition CH (theoretical graphene [36] (Figure 4)) and experimental graphene [37] ; 2) theoretical single-side hydrogenated graphene of composition CH; 3) theoretical single-side hydrogenated graphene of composition C_2H (graphene); 4) experimental hydrogenated epitaxial graphene, bilayer graphene and a few layer graphene on SiO_2 or other substrates; 5) experimental and theoretical single-external side hydrogenated single-walled carbon nanotubes, and experimental hydrofullerene $C_{60}H_{36}$; 6) experimental single-internal side hydrogenated (up to C_2H or CH composition) graphene nanoblisters with intercalated high pressure H_2 gas inside them, formed on a surface of highly oriented pyrolytic graphite or epitaxial graphene under the atomic hydrogen treatment; and 7) experimental hydrogenated graphite nanofibers—multigraphene with intercalated solid H_2 nanoregions of high density inside them [2] [3] , relevant to solving the current problem of the hydrogen storage in fuel-cell-powered vehicles and other clean energy applications. It is expedient to note [2] [3] that there are a number of fundamental open questions in this field to be further studied.

The Figure 4 shows the diamond-like distorted hexagonal network with carbon in sp^3 hybridization [36] , the carbon atoms are shown in gray and the hydrogen atoms in white.

3.2. Determination of thermodynamic characteristics of graphene hydrides

Some theoretical, experimental and semi-empirical results [3] [36] -[41] of determination of thermodynamic characteristics of graphene hydrides, including determination of the energy of formation of graphene hydrides from graphene and atomic gaseous hydrogen, the energy of breaking-down of C-H sp^3-bonds (ΔH (C-H)) in graphene hydrides, the energy of formation of graphene hydrides from atomic gaseous carbon and atomic gaseous hydrogen, the energy of breaking-down of C-C sp^3-bonds (ΔH (C-C)) in graphene hydrides, and some other characteristics are considered (Table 1) from [3] .

The critical comparison of the theoretical and experimental quantities, resulting in determination of the errors of the theoretical evaluations, are also included in [3] . In the light of the analytical results [3] , two open questions formulated by K.S. Novoselov, A.K. Geim et al. (2009) [37] are considered, namely as:

1) Is the theoretical graphene (CH) [36] "the until-now-theoretical material"?

2) May the experimental hydrogenated graphene (CH) [37] have "a more complex hydrogen bonding than the one suggested by theory [36] "?

Two other fundamental open questions [3] are also discussed, namely as:

3) Can hydrogenated graphene layers nanostructures (graphene hydrides) with a high energy of C-C sp^3-bonds close to that for graphene exist?

4) Can a solid (or liquid) molecular hydrogen nanophase be intercalated into hydrogenated graphite nanofibers.

Figure 4. The theoretical graphene structure in chair configuration.

Table 1. Analytical and other characteristics.

	Value/quantity (analytical: an.)				
	ΔH (C-H) (eV)	ΔH (bind.) (eV)	ΔH (C-C) (eV)	ΔH (des.) (eV)	K^0 (des.) (s^{-1})
Graphene [36]	(2.5 ± 0.1) an.	6.56	(2.7) an.	(2.5) an.	
Graphene [39]	1.50	5.03	(2.35) an.	(1.5) an.	
Graphene [38]	2.46 ± 0.17			2.46 ± 0.17	$(2.1 \pm 0.5) \times 10^{17}$
Graphene [38] an.	2.46 ± 0.17			2.46 ± 0.17	2.0×10^{15}
Graphene membrane [37] an.	2.5 ± 0.1 2.6 ± 0.1			2.5 ± 0.1 2.6 ± 0.1	7×10^{12} 5×10^{13}
Graphene epitaxial [37] an.	1.84 1.94			1.84 1.94	7×10^{12} 5×10^{13}
Graphene epitaxial, TDS #1 [37] an.				0.6 ± 0.3	2×10^{7}
Graphene epitaxial, TDS #2 [37] an.				0.6 ± 0.3	1×10^{6}
Graphene epitaxial, TDS #3 [37] an.				0.23 ± 0.05	2.4
Graphene [39]		7.40	(4.93) an.		
Graphene* [31]-[33]	2.6	9.95	4.9	2.6	
Graphite [31]-[33]		7.41 ± 0.05	4.94 ± 0.03		
Diamond [31]-[33]		7.38 ± 0.04	3.69 ± 0.02		
Hydrofullerene $C_{60}H_{36}$ [40]	2.64 ± 0.01				
Hydrogenated carbon nanotubes (C_2H, [41])	2.5 ± 0.2				

3.3. A treatment & interpretation of some recent STM, STS, HREELS/LEED, PES, ARPS and Raman spectroscopy data on Hydrogen sorption with epitaxial grapheme

Herein, some results of the thermodynamic analysis [3] of the data [42] -[46] on hydrogen sorption with epitaxial graphene are presented in Table 2, Table 3. In the framework of the formal kinetics approximation of the first order rate reaction, the characteristic quantities for the processes [42] -[46] of hydrogen sorption (the rate constant (K (des.) = 1/0.63 (des.)), the activation energy (ΔH (des.)), the pre-exponential factor of the rate constant (K^0 (des.)) are determined.

Evaluation of the HREELS elastic peak FWHM of hydrogenated graphene on SiC substrate (SiC-D/ QFMLG-H) upon annealing is shown in Figure 5, where the uncertain annealing temperature is estimated to be 5%, error bars represent the σ variation of FWHM measured across the entire surface of several samples [44] .

Some models (Figure 6) and characteristics [34] [35] of hydrogen chemisorption on graphite (on the basal and edge planes) have been used for interpretation of the results of data [42] - [46] treatment, in relation with the hydrogen storage problem. Schemes of some theoretical models of chemisorption of atomic hydrogen on graphite on the basal and edge planes [34] [35] are presented in Figure 6.

As it is shown in [2] [3] , there are a number of fundamental open questions also in this field to be further studied.

Table 2. Analytical results (from [3]).

Materials	Values/quantities		
	ΔH (des.) (eV)	K^0 (des.) (s^{-1})	τ 0.63 (des.) 553 K (s)
Graphene flakes/SiO_2 [42]	0.11 ± 0.07	0.15	70
Graphene/Ni [43]			130
HOPG [43]			130
(SiC-D/QFMLG-H) [44]	0.7 ± 0.2	9×10^2	
(SiC-D/QFMLG) [44]	2.0 ± 0.6	1×10^6	

Table 3. Some other analytical results (from [3]).

Samples	Values/quantities (analytical: an.)			
	ΔH (des.) I (eV)	K^0 (des.) I (s^{-1})	ΔH (des.) II (eV)	K^0 (des.) II (s^{-1})
1LG-15W (grapheme) [45]	0.6 ± 0.2	2×10^4	0.19 ± 0.07	3×10^{-2}
2LG-15W (bi-graphene) [45]			0.9 ± 0.3	1×10^3
1LG-5W (grapheme) [45]	0.15 ± 0.04	2×10^{-2}	0.31 ± 0.07	5×10^{-1}
2LG-5W (bi-graphene) [45]	0.50 ± 0.15	2×10^5	0.40 ± 0.15	1
HOPG [47], TDS-peaks I, II	0.6 ± 0.2	1.5×10^4	1.0 ± 0.3	2×10^6
Graphene/SiC [46]			3.6	2×10^{14}
HOPG [48], TDS-peaks I, II	2.4 [48]		4.1 [48]	
HOPG [48], TDS-peak I	(2.4 ± 0.5) an.	(2×10^{10}) an.		
GNF [5] [11], TDS-peaks I, II	(2.4 ± 0.5) an.			

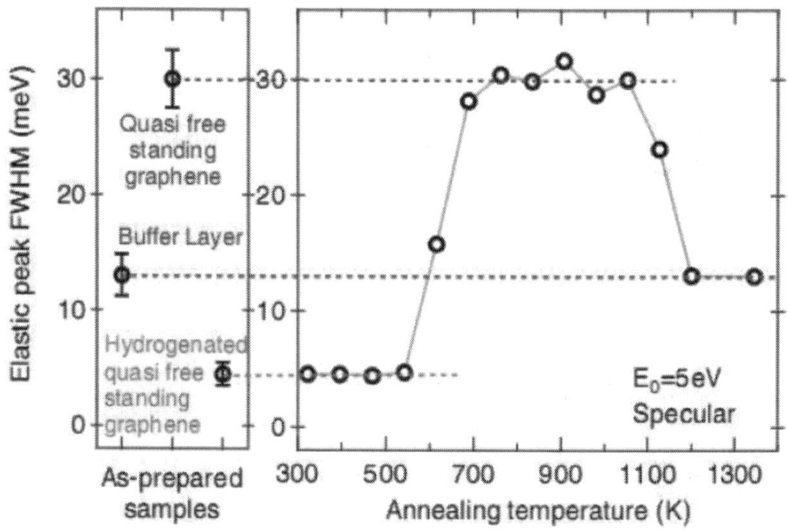

Figure 5. Evaluation of the HREELS elastic peak FWHM of hydrogenated graphene on SiC substrate upon annealing.

3.4. On the physics of intercalation of hydrogen into surface graphene-like nanoblisters in pyrolytic graphite and epitaxial grapheme

In the light of the analytical results [3] , the physics of the spontaneous process ([47] - [50] and others) of intercalation of gaseous molecular hydrogen of a high pressure into nanoblisters—the surface nano-size bulges with monolayer graphene-like "walls", those being formed (at definite conditions of hydrogenation in atomic gaseous hydrogen) on a surface of highly oriented pyrolytic graphite (Figure 7, Figure 8) and epitaxial graphene (Figure 9, Figure 10). A fundamental open question [3] [51] namely as “Can gaseous atomic hydrogen penetrate through a perfect graphene monolayer?” is also discussed.

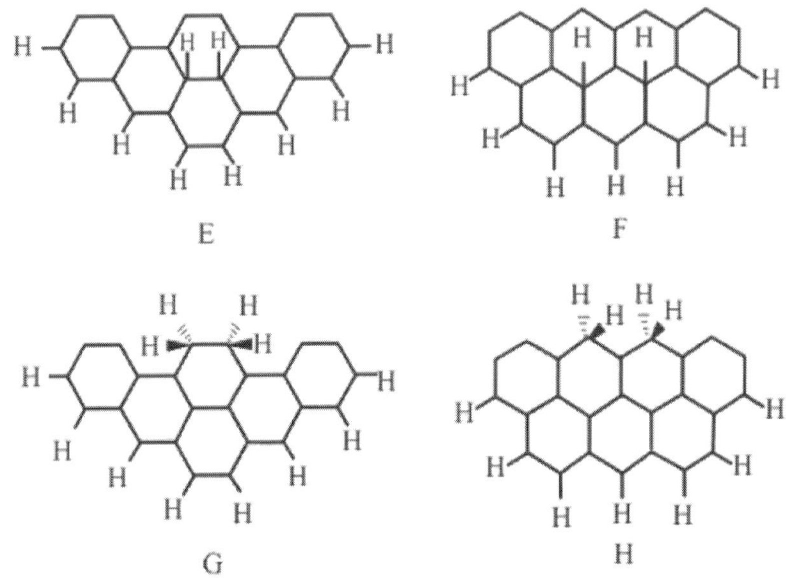

Figure 6. Schemes of some theoretical models of chemisorption of atomic hydrogen on graphite on the basal and edge planes.

On Figure 7 there are STM images of the untreated HOPG sample [49] (under ambient conditions) taken from areas of (a) 60.8 × 60.8 nm and (b) 10.9 × 10.9 nm (high resolution image of the square in image (a)). (c). AFM image (area of 1 × 1 nm) of the HOPG sample subjected to atomic hydrogen dose (D) of 1.8 × 10^{16} H^0/cm^2. (d) Surface height profile obtained from the AFM image reported in (c). The STM tunnel V_{bias} and current are 50 - 100 mV and 1 - 1.5 mA, respectively.

Figure 8 shows a hydrogen storage efficiency of HOPG samples [49] , desorbed molecular hydrogen (Q) versus dose (D) of atomic hydrogen exposure (Figure 8(a)) and STM image for 600 × 600 nm area of the HOPG sample subjected to atomic hydrogen dose of 1.8 × 10^{16} H^0/cm^2, followed by hydrogen thermal desorption (Figure 8(b)).

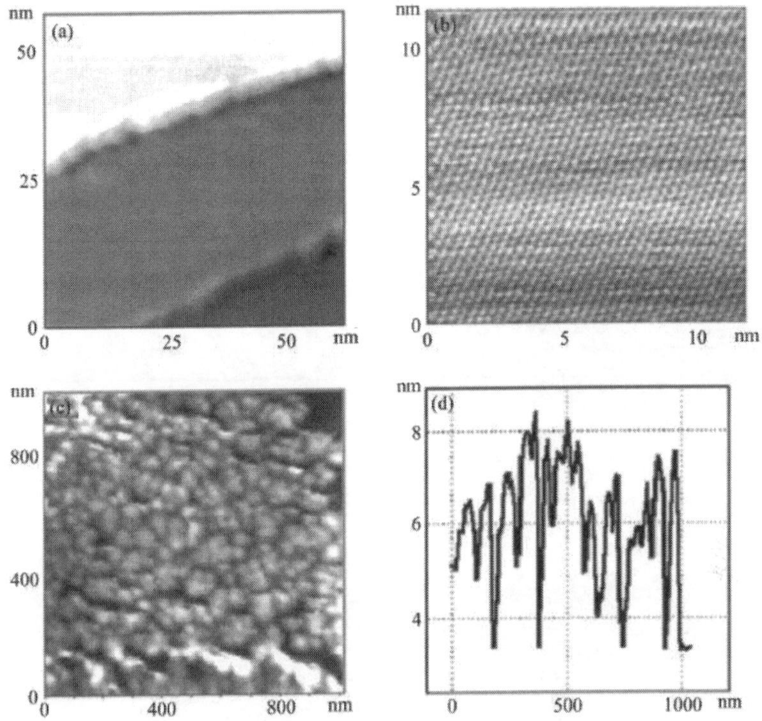

Figure 7. High resolution STM images of the untreated HOPG sample taken from varios areas of: (a) 60.8 × 60.8 nm; (b) 10.9 × 10.9 nm; (c) AFM image (area of 1 × 1 nm); (d) Surface height profile obtained from the AFM image reported in (c).

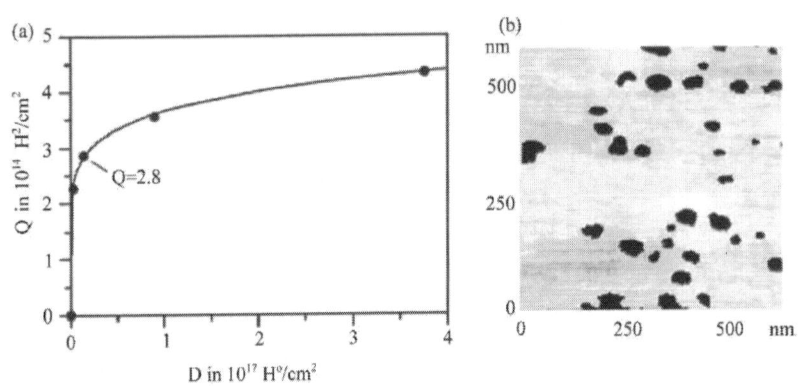

Figure 8. (a) Desorbed molecular hydrogen (Q) versus dose (D) of atomic hydrogen exposure; (b) STM image for 600 × 600 nm area of the HOPG sample subjected to atomic hydrogen dose followed by hydrogen thermal desorption.

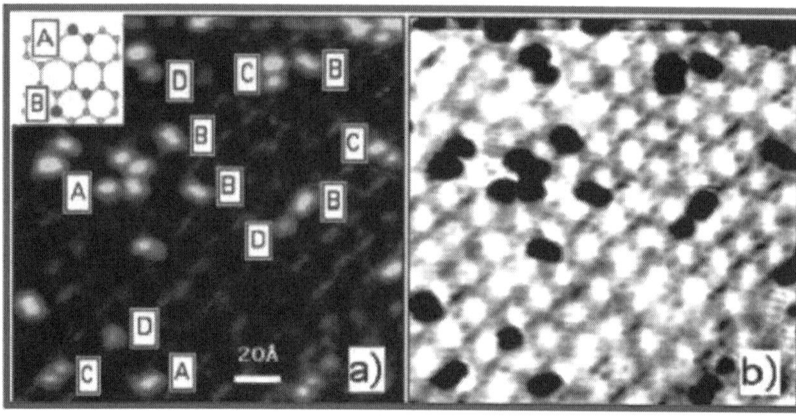

Figure 9. (a) STM image of hydrogenated graphene; (b) The same image as in (a) with inverted color scheme.

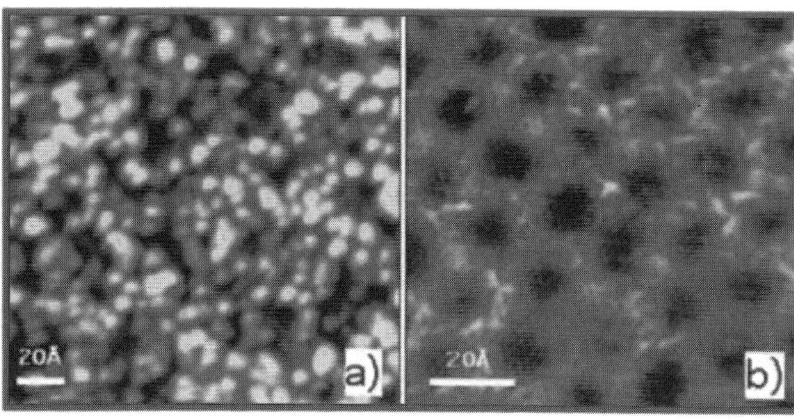

Figure 10. (a) STM image of the graphene surface after extended hydrogen exposure; (b) Large graphene area recovered from hydrogenation by annealing to 1073 K.

Figure 9(a) presents a scanning tunneling microscopy (STM) image of hydrogenated graphene [50] . The bright protrusions visible in the image are atomic hydrogen adsorbate structures identified as A = ortho-dimers, B = para-dimers, C = elongated dimers, D = monomers (imaging parameters: V_t= −0.245 V_t, I_t = −0.26 nA). There is inset in (a): ortho- and para-dimer configuration on the graphene lattice (A and B, respectively). Figure 9(b) shows the same image as in Figure 9(a) but with inverted color scheme, giving emphasis to preferential hydrogen adsorption along the 6 × 6 modulation on the SiC (0001)—(1 × 10 surface). Hydrogen dose at T_{beam} = 1600 K, t = 5 s, F = 10^{12} - 10^{13} atoms/cm^2·s.

Figure 10(a) presents the STM image of the graphene surface after extended hydrogen exposure [50] . The bright protrusions visible in the image are atomic hydrogen clusters (imaging parameters: V_t = −0.36 V, I_t = −0.32 nA). Hydrogen dose at T = 1600 K, t = 90 s, F = 10^{12} - 10^{13}atoms/cm^2·s. Figure 10(b) exibit a large graphene area recovered from hydrogenation by annealing to 1073 K (imaging parameters: V_t = −0.38 V, I_t = −0.41 nA).

Authors of [51] noted that their test calculations show that the barrier for the penetration of a hydrogen atom through the six-membered ring of graphene is larger than 2.0 eV. Thus, they believe that it is almost impossible for a hydrogen atom to pass through the six-membered ring of graphene at room temperature (from a private communication with H.G. Xiang and M.-H. Whangbo). On the other hand, in the light of analysis [3] of a number of recent experimental studies, there is a real possibility that a hydrogen atom can pass through the graphene network at room temperature. This is the case of existing relevant defects in graphene, i.e., grain boundaries (particularly, the triple junctions of them) and/or vacancies.

3.5. On the physics of the elastic & plastic deformation of graphene (graphene-like) walls in hydrogenated graphite nanofibers

Herein, some results of the thermodynamic analysis [2] [3] of the unique (extraordinary) data [4] - [12] are considered. Mechanical behavior (the elastic and plastic deformation (Figure 11, Figure 12)) of the graphene (graphene-like) walls in graphite nanofibers (GNFs) under definite hydrogenation conditions are analyzed and interpreted, relevance to the hydrogen storage problem. There are a number of fundamental open questions also in this field to be further studied. Here,Figure 11 shows an increase in GNF width due to H_2O adsorption [5] .

Figure 12 shows the TEM of the dehydrogenated GNFs, by [11] .

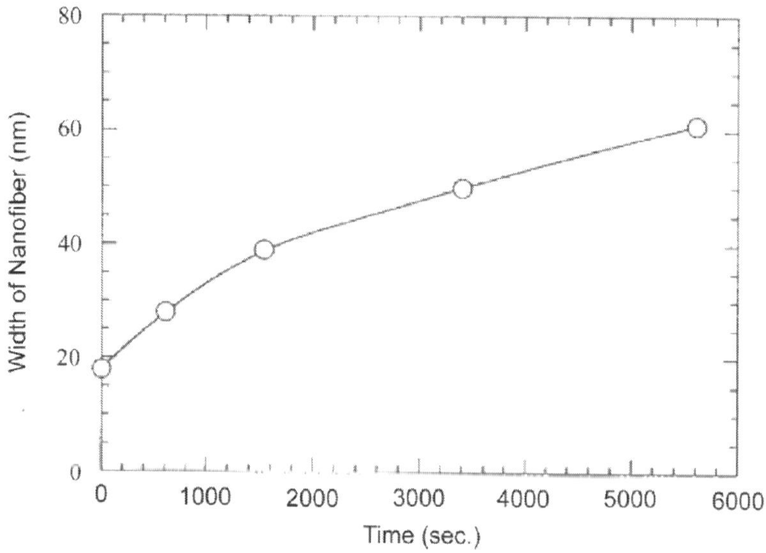

Figure 11. Increase in GNF width due to H_2O adsorption.

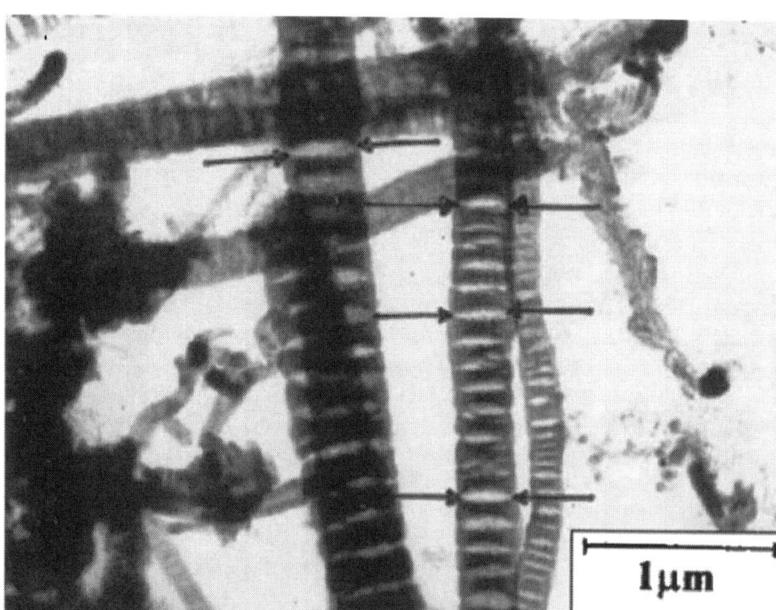

Figure 12. TEM of the dehydrogenated GNF.

3.6. On the physics of engineering of "super" hydrogen storage Carbonaceous nanomaterials, in the light of analysis of the Rodriguez-Baker extraordinary data and some others

Herein, in the light of a constructive critical analysis (including the analytical results [2] [3]) of a complete set of the extraordinary experimental (nobody reproduced up to nowadays) data [4] - [12] and some related ones [52] [53], the physical fundamentals of developing of a key breakthrough technology of a compact and energy efficient hydrogen storage in hydrogenated graphite nanofibers (Figure 12, 1, 2) are considered. As far as we know, both authors [4] - [8] (1998-2005) and authors [9] - [12] (2000-2006) have never crossed out their extraordinary non-reproduced experimental results. On the other hand, article [4] has been cited (within 1998-2013) 168 times in Scopus, and article [5] —196 times. But, as far as we know, nobody has analyzed and interpreted a complete set of [4] - [12] data. As is shown in the present study, the most recent negative test-results [21] [23] , with respect to [4] - [12] data can be caused by using in [21] [22] the non-adequate GNF samples (including samples and regimes supplied for this test by Rodriguez and Baker themselves). Some ethical aspects of this long-term situation are discussed, as well.

4. CONCLUSIONS

1) As is shown in this work, in the light of analysis [2] [3] , there are a number of fundamental open questions, relevance to developing of a key breakthrough technology of a compact and energy efficient hydrogen storage in hydrogenated graphite nanofibers (Figure 1, Figure 2), to be further studied.

2) These fundamental open questions [2] [3] [31] - [33] may be solved within several years, i.e., within the necessary time frame (Figure 3).

3) A constructive open discussion and, may be, a related international cooperation on solving the above considered open questions seems expedient, relevant to the promotion of further developments, particularly in the developing of basic grounds of a possible break-through in nanotechnology for hydrogen on-board storage and other clean energy applications [2] [3] .

ACKNOWLEDGEMENTS

This work has been supported by the RFBR (Project #14-08-91376 CT) and the TUBITAK (Project # 213M523).

REFERENCES

1 Zuettel, A. (2011) Hydrogen the Future Energy Carrier. In: Materials of International Hydrogen Research Showcase 2011, University of Birmingham, Birmingham, 13-15 April 2011. http://www.uk-shec.org.uk/ukshec/showcase/ShowcasePresentations.html

2 Nechaev, Y.S. and Veziroglu, T.N. (2013) On Thermodynamic Stability of Hydrogenated Graphene Layers, Relevance to the Hydrogen On-Board Storage. The Open Fuel Cells Journal, 6, 21-39.

3 Nechaev, Y.S. and Veziroglu, T.N. (2013) Thermodynamic Aspects of the Stability of the Graphene/Graphane/Hy- drogen Systems, Layers, Relevance to the Hydrogen On-Board Storage Problem. Advances in Materials Physics and Chemistry, 3, 255-280.

4 Chambers, A., Park, C., Terry, R., Baker, K. and Rodriguez, N.M. (1998) Hydrogen Storage in Graphite Nanofibers. Journal of Physical Chemistry B, 102, 4253-4256.

5 Park, C., Anderson, P.E., Chambers, A., Tan, C.D., Hidalgo, R. and Rodriguez, N.M. (1999) Further Studies of the Interaction of Hydrogen with Graphite Nanofibers. Journal of Physical Chemistry B, 103, 10572-10581.

6 Nelly, M., Rodriguez, N.M., Terry, R. and Baker, K. (1997) Storage of Hydrogen in Layered Nanostructures. US Patent No. 5653951.

7 Nelly, M., Rodriguez, N.M., Terry, R. and Baker, K. (2000) Method for Introducing Hydrogen into Layered Nanostructures. US Patent No. 6159538.

8 Baker, R.T.K. (2005) Encyclopedia of Materials: Science and Technology. Elsevier, Amsterdam, 932.

9 Gupta, B.P. and Srivastava, O.N. (2000) Synthesis and Hydrogenation Behavior of Graphitic Nanofibers. Journal of Hydrogen Energy, 25, 825-830.

10 Gupta, B.P. and Srivastava, O.N. (2001) Further Studies on Microstructural Characterization and Hydrogenation Behavior of

Graphitic Nanofibers. International Journal of Hydrogen Energy, 26, 857-862.

11 Gupta, B.K., Tiwari, R.S. and Srivastava, O.N. (2004) Studies on Synthesis and Hydrogenation Behavior of Graphitic Nanofibers Prepared through Palladium Catalyst Assisted Thermal Cracking of Acetylene. Journal of Alloys and Com- pounds, 381, 301-308.

12 Gupta, B.P. and Srivastava, O.N. (2006) New Carbon Variants: Graphitic Nanofibers (Nano-Springs, Nano-Shoeckers) as Hydrogen Storage Materials. International Scientific Journal for Alternative Energy and Ecology, 5, 63.

13 Suh, M.P., Park, H.J., Prasad, T.K. and Lim, D.W. (2012) Hydrogen Storage in Metal-Organic Frameworks. Chemical Reviews, 112, 782-835.

14 Tozzini, V. and Pellegrini, V. (2013) Prospects for Hydrogen Storage in Graphene. Physical Chemistry Chemical Physics, 15, 80-89.

15 Satyapal, S., Petrovic, J., Read, C., Thomas, G. and Ordaz, G. (2007) The U.S. Department of Energy's National Hydrogen Storage Project: Progress towards Meeting Hydrogen-Powered Vehicle Requirements. Catalysis Today, 120, 246-256.

16 DOE (2012) Targets for Onboard Hydrogen Storage Systems for Light-Duty Vehicles.http://wwwl.eere.energy.gov/ hydrogenandfuelcells/ storage/%20pdfs/targets_onboard_hydro_storage.pdf

17 Fruchart, D. (2011) Large Scale Development of Metal Hydrides for Stationary and Nomad Hydrogen Storage Units. Which Are Potential Partners to Built Innovative Solutions for Sustainable and Clean Energy Systems? In: Materials of International Hydrogen Research Showcase 2011, University of Birmingham, Birmingham, 13-15 April 2011.http://www.uk-shec.org.uk/uk-shec/ showcase/ Showcase Presentations.html

18 Akiba, E. (2011) Hydrogen Related R&D and Hydrogen Storage Materials in Japan. In: Materials of International Hydrogen Research Showcase 2011, University of Birmingham, Birmingham, 13-15 April 2011.

19 Kim, J.W. (2011) Current Status of R&D on Hydrogen Production and Storage in Korea. In: Materials of International Hydrogen Research Showcase 2011, University of Birmingham, Birmingham, 13-15 April 2011.

20 Maeland, A.J. (2002) The Storage of Hydrogen for Vehicular Use—A Review and Reality Check. International Scientific Journal for Alternative Energy and Ecology, 1, 19-29.

21 Rzepka, M., Bauer, E., Reichenauer, G., Schliermann, T., Bernhardt, B., Bohmhammel, K., Henneberg, E., Knoll, U., Maneck, H.E. and Braue, W. (2005) Hydrogen Storage Capacity of Catalytically Grown Carbon Nanofibers. The Journal of Physical Chemistry B, 109, 14979-14989.

22 Reichenauer, G., Rzepka, M., Bauer, E., Schliermann, T., Bernhardt, B., Bohmhammel, K., Henneberg, E., Knoll, U., Maneck, H.E. and Braue, W. (2006) Hydrogen Storage Capacity of Catalytically Grown Carbon Nanofibers. Proceedings of International Conference on Carbon, The Robert Gordon University, Aberdeen, 6-21 July 2006.

23 Ramos, A., Cameán, I. and García, A.B. (2013) Graphitization Thermal Treatment of Carbon Nanofibers. Carbon, 59, 2-32.

24 Camacho, R.M. and Guirado-López, R.A. (2013) Adsorption and Diffusion of Hydrogen on C_{60}-Supported Pt_n Clusters. The Journal of Physical Chemistry C, 117, 10059-10069.

25 Gao, Q., Qu, F.Y., Lin, H.M. and Zheng, W.T. (2013) A Simple Method to Synthesize Graphitic Mesoporous Carbon Materials with Different Structures. Journal of Porous Materials, 20, 983-988.

26 He, Z., Wang, S., Iqbal, Z. and Wang, X.Q. (2013) Hydrogen Storage in Hierarchical Nanoporous Silicon-Carbon Nanotube Architectures. International Journal of Energy Research, 37, 754-760.

27 Lueking, A.D., Yang, R.T., Rodriguez, N.M. and Baker, R.T.K. (2004) Hydrogen Storage in Graphite Nanofibers: Effect of Synthesis Catalyst and Pretreatment Conditions. Langmuir, 20, 704-721.

28 Ströbel, R., Garche, J., Moseley, P.T., Jörissen, L. and Wolf, G. (2006) Hydrogen Storage by Carbon Materials. Journal of Power Sources, 159, 781-801.

29 Yürüm, Y., Taralp, A. and Veziroglu, T.N. (2009) Storage of Hydrogen in Nanostructured Carbon Materials. International Journal of Hydrogen Energy, 34, 3784-3798.

30 Kowalczyk, P., Holyst, R., Terrones, M. and Terrones, H. (2007) Hydrogen Storage in Nanoporous Carbon Materials; Myth and Facts. Physical Chemistry Chemical Physics, 9, 1786-1792.

31 Nechaev, Y.S. (2012) Solid Hydrogen in Multigraphane Nanostructures. International Scientific Journal for Fundamental and Applied Physics, 1, 38-60.

32 Nechaev, Y.S. (2011) On the Solid Hydrogen Carrier Intercalation in Graphane-Like Regions in Carbon-Based Nanostructures. International Journal of Hydrogen Energy, 36, 9023-9031.

33 Nechaev, Y.S. (2011) The High-Density Hydrogen Carrier Intercalation in Graphane-Like Nanostructures, Relevance to Its On-Board Storage in Fuel-Cell-Powered Vehicles. The Open Fuel Cells Journal, 4, 16-29.

34 Nechaev, Y.S. (2006) The Nature, Kinetics, and Ultimate Storage Capacity of Hydrogen Storage by Carbon Nanostructures. Physics-Uspekhi, 49, 563-591.

35 Nechaev, Y.S. (2010) Carbon Nanomaterials, Relevance to the Hydrogen Storage Problem. Journal of Nano Research, 12, 1-44.

36 Sofo, J.O., Chaudhari, A.S. and Barber, G.D. (2007) Graphane: A Two-Dimensional Hydrocarbon. Physical Review B, 75, Article ID: 153401.

37 Elias, D.C., Nair, R.R., Mohiuddin, T.M.G., Morozov, S.V., Blake, P., Halsall, M.P., Ferrari, A.C., Boukhvalov, D.W., Katsnelson, M.I., Geim, A.K. and Novoselov, K.S. (2009) Control of Graphene's Properties by Reversible Hydrogenation: Evidence for Graphane. Science, 323, 610-613.

38 Openov, L.A. and Podlivaev, A.I. (2010) Thermal Desorption of Hydrogen from Graphane. Technical Physics Letters, 36, 31-33.

39 Dzhurakhalov, A.A. and Peeters, F.M. (2011) Structure and Energetics of Hydrogen Chemisorbed on a Single Graphene Layer to Produce Graphane. Carbon, 49, 3258-3266.

40 Pimenova, S.M., Melkhanova, S.V., Kolesov, V.P. and Lobach, A.S. (2002) The Enthalpy of Formation and C-H Bond Enthalpy Hydrofullerene $C_{60}H_{36}$. The Journal of Physical Chemistry B, 106, 2127-2130.

41 Bauschlicher Jr., C.W. and So, C.R. (2002) High Coverages of Hydrogen on (10.0), (9.0) and (5.5) Carbon Nanotubes. Nano Letters, 2, 337-341.

42 Wojtaszek, M., Tombros, N., Garreta, A., Van Loosdrecht, P.H.M and Van Wees, B.J. (2011) A Road to Hydrogenating Graphene by a Reactive Ion Etching Plasma. Journal of Applied Physics, 110, Article ID: 063715.

43 Castellanos-Gomez, A., Wojtaszek, M., Arramel, Tombros, N. and Van Wees, B.J. (2012) Reversible Hydrogenation and Bandgap Opening of Graphene and Graphite Surfaces Probed by Scanning Tunneling Spectroscopy. Small, 8, 1607-1613.

44 Bocquet, F.C., Bisson, R., Themlin, J.M., Layet, J.M. and Angot, T. (2012) Reversible Hydrogenation of Deuterium- Intercalated Quasi-Free-Standing Graphene on SiC (0001). Physical Review B—Condensed Matter and Materials Physics, 85, Article ID: 201401.

45 Luo, Z.Q., Yu, T., Kim, K.J., Ni, Z.H., You, Y.M., Lim, S., Shen, Z.X., Wang, S.Z. and Lin, J.Y. (2009) Thickness- Dependent Reversible Hydrogenation of Graphene Layers. ACS Nano, 3, 1781-1788.

46 Watcharinyanon, S., Virojanadara, C., Osiecki, J.R., Zakharov, A.A, Yakimova, R., Uhrberg, R.I.G. and Johansson, L.I. (2011) Hydrogen Intercalation of Graphene Grown on 6H-SiC (0001). Surface Science, 605, 1662-1668.

47 Hornekaer, L., Šljivančanin, Ž., Xu, W., Otero, R., Rauls, E., Stensgaard, I., Lægsgaard, E., Hammer, B. and Besenbacher, F. (2006) Metastable Structures and Recombination Pathways for Atomic Hydrogen on the Graphite (0001) Surface. Physical Review Letters, 96, Article ID: 156104.

48 Waqar, Z., Klusek, Z., Denisov, E., Kompaniets, T., Makarenko, I., Titkov, A. and Saleem, A. (2000) Effect of Atomic Hydrogen Sorption and Desorption on Topography and Electronic Properties of Pyrolytic Graphite. Electrochemical Society Proceedings, 16, 254-265.

49 Waqar, Z. (2007) Hydrogen Accumulation in Graphite and Etching of Graphite on Hydrogen Desorption. Journal of Materials Science, 42, 1169-1176.

50 Balog, R., Jørgensen, B., Wells, J., Lægsgaard, E., Hofmann, P., Besenbacher, F. and Hornekær, L. (2009) Atomic Hydrogen Adsorbate Structures on Graphene. Journal of the American Chemical Society, 131, 8744-8745.

51 Xiang, H.J., Kan, E.J., Wei, S.H., Gong, X.G. and Whangbo, M.H. (2010) Thermodynamically Stable Single-Side Hydrogenated Graphene. Physical Review B, 82, Article ID: 165425.

CHAPTER 9

Employment of Two-Stage Oxygen Feeding to Control Temperature in a Downdraft Entrained-Flow Coal Gasifier

Ting Wang[1], Xijia Lu[1], Heng-Wen Hsu[2], Cheng-Hsien Shen[2]

[1]Energy Conversion & Conservation Center, University of New Orleans, New Orleans, LA, USA

[2]Green Energy and Environment Research Laboratory, Industrial Technology Research Institute, Hsinchu City, Taiwan

ABSTRACT

The traditional practice of employing a two-stage coal-fed gasification process is to feed all of the oxygen to provide a vigorous amount of combustion in the first stage but only feed the coal without oxygen in the second stage to allow the endothermic gasification process to occur downstream of the second stage. One of the merits of this 2-stage practice is to keep the gasifier temperature low downstream from the 2nd stage. This helps to extend the life of refractory bricks, decrease gasifier shut-down frequency for scheduled maintenance, and reduce the maintenance costs. In this traditional 2-stage practice, the temperature reduction in the second stage is achieved at the expense of a higher than normal temperature in the first stage. This study investigates a concept totally opposite to the traditional two-stage coal feeding practices in which the injected oxygen is split between the two stages, while all the coal is fed into the first stage. The hypothesis of this two-stage oxygen injection is that a distributed oxygen injection scheme can also distribute the release of heat to a larger gasifier volume and, thus, reduce the peak temperature distribution in the gasifier. The increased life expectancy and reduced maintenance of the refractory bricks can prevail in the entire gasifier and not just downstream from the second stage. In this study, both experiments and computational simulations have been performed to verify the hypothesis. A series of experiments conducted at 2.5 - 3.0 bars shows that the peak temperature and temperature range in the gasifier do decrease from 600°C - 1550°C with one stage oxygen injection to 950°C - 1230°C with a 60 - 40 oxygen split-injection. The CFD results conducted at 2.5 bars show that 1) the carbon conversion ratio for different oxygen injection schemes are all above 95%; 2) H_2 (about 70% vol.) dominates the syngas composition at the exit; 3) the 80% - 20% case yields the lowest peak temperature and the most uniform temperature distribution along the gasifier; and 4) the 40% - 60% case produces the syngas with the highest HHV. Both experimental data and CFD predictions verify the hypothesis that it is feasible to reduce the peak temperature and achieve more uniform temperature in the gasifier by adequately controlling a two-stage oxygen injection with only minor changes of the composition and heating value of the syngas.

KEYWORDS

Two-Stage Oxygen Feeding, Entrained-Flow Coal Gasification, Gasification Simulation, Syngas Composition, Clean Coal Technology, Multiphase Flow Simulation

1. INTRODUCTION

1.1. Brief Review of Fuel Feeding Scheme of Entrained Coal Gasification

Gasification is an incomplete oxidation process converting various carbon-based feedstocks into clean synthetic gas (syngas), which is primarily a mixture of hydrogen (H_2) and carbon monoxide (CO) with minor amounts of methane (CH_4) and inert nitrogen gas (N_2). There are four main classes of gasifiers: fixed bed gasifier, fluidized bed gasifier, entrained flow gasifier, and transport gasifier. The fixed bed gasifier and fluidized bed gasifier have low (425°C - 650°C) and moderate (900°C - 1050°C) outlet gas temperatures, respectively. Entrained flow gasifiers have high outlet temperatures (1250°C - 1600°C) and operate in the slagging range (the ash is fully liquid with low viscosity). The main benefits of entrained flow gasifiers can be summarized as: the ability to handle practically any coal as feed, syngas being free of oils and tars, high carbon conversion, suitable for synthesis gas products, and high throughput because of high reaction rates at elevated temperature.

According to the direction of the main flow inside the gasifiers, the entrained flow gasifiers can be classified as up-flow design such as the Shell SCGP and E-gas processes, and down-flow design such as the GE and Simmens gasifiers. For down-flow coal gasifier (GE gasifier), coal slurry and oxidant are introduced from the top of the gasifier. The hot raw gas produced during the gasification process flows toward the bottom of the gasifier. The ash forming components melt in the gasifier, flow down the walls of the reactor, and finally leave the reactor as a liquid slag.

Based on the fuel injection scheme, the entrained flow gasifiers can be categorized as single-stage injection and two-stage injection. The E-Gas process of ConocoPhillips features a two-stage up-flow gasifier design consisting of a horizontal cylindrical vessel as the first stage and a vertical cylindrical vessel as the second stage. Approximately 78% of the coal slurry and 100% of the oxygen are fed in the first stage. Additional coal-water slurry without oxygen is injected into this hot syngas in the second gasifier stage. Endothermic gasification reactions occur between the hot syngas and the second stage coal feed. This lowers the temperature of the syngas and increases the cold gas efficiency of the process. One of the most important advantages of this 2-stage practice is to keep the gasifier temperature low downstream from the 2nd stage and thus extends the life of the refractory bricks, decreases gasifier shut-down frequency for scheduled maintenance, and reduces maintenance costs. Although local stoichiometry as well as dynamic variations in the local stoichiometry will also affect corrosion of the refractory liner, this study only focuses on managing the temperature distribution.

1.2. Motivation, Hypothesis, and Objectives

Contrast to the traditional concept of two-stage coal feeding practices of injecting coal slurry into two different stages, it is hypothesized that splitting oxygen between two stages, while feeding all the coal in the first stage, i.e., a distributed oxygen injection scheme, can distribute heat release into a larger gasifier's volume and hence reduces the peak temperature and make temperature distribution more uniform in the gasifier. In this approach, the increased life expectancy and reduced maintenance of the refractory can prevail in the entire gasifier-not just downstream from the second stage as in the traditional two-stage coal slurry fed system. To verify this hypothesis, both experiments and computational simulations are conducted. Since conducting experiments are expensive, time consuming, and usually performed with limited cases, it would be more effective and meaningful to establish a computational fluid dynamics (CFD) model to simulate effects of varying operating conditions to help obtain a preliminary understanding of the fundamental thermal-flow physics inside the gasifier. However, since the coal gasification is a very complicated multi-component and multi-phase process, a trustful and fully calibrated computational model is not available. To obtain a usable computational model as a meaningful tool, the experimental data is utilized to guide interpretation of the computational result in low operating pressure (2.5 bars) cases. Afterwards, the computational model is employed to perform detailed thermal-flow and chemical reaction simulation for more parametric studies that have not been performed by experiments. To this end, the objectives of this paper are:

1) Conduct experiments to investigate the effect of a controlled two-stage distributed oxygen approach on a gasifier's inner wall temperature distribution.

2) Interpret the CFD results and determine how the results can be effectively used by comparing them with the baseline experimental data.

3) Use the CFD model to simulate experimental cases with different oxygen split fed from 100% to 40% under the condition of 2.5 bars and 0.6 stoichiometric ratio.

4) Investigate computationally the effect of distributed oxygen fed on gasification performance and temperature distribution under a higher operating pressure (25 bars) and 0.4 stoichiometric ratio..

2. EXPERIMENTAL FACILITY

Industrial Technology Research Institute (ITRI) has designed a 2 ton/day, two-stage coal slurry fed, down draft entrained-flow gasifier for this study, as shown in Figure 1. This experimental facility is located in the southern Taiwanese city of Kaohsiung. The height of the gasification section above the quench section is 4.831 m, the inner diameter is 0.27 m, and the thickness of the refractory is 0.35 m. The coal powder is first mixed with water at a mass ratio of approximately 43% (coal)/57% (water). The coal slurry is then transported through the center tube of the injector as shown inFigure 2. All the oxygen is fed through the outer concentric annular pipe and is mixed with the slurry coal before injection. The total oxygen fed to the gasifier is about 60% of the stoichiometric amount needed for complete coal combustion (i.e., with a stoichiometric ratio of 0.6). The injector is water cooled through an annular water jacket wrapping around the fuel and oxygen passages. The blunt injector tip with multiple holes is derived from a previous study which investigated the performance and life expectancy of two injectors' designs [1] . It was discovered that the injector with a blunt tip geometry was able to reduce temperature gradient along the injector and significantly increase the life of the fuel injector more than the injector with the conical tip. The injector used in this study is designed and manufactured for experimental use only with three options for feeding oxygen-via the outer eight holes or inner four holes for external mixing, or using premixing scheme as shown in Figure 2. In this study, premixing scheme is used and other holes are plugged.

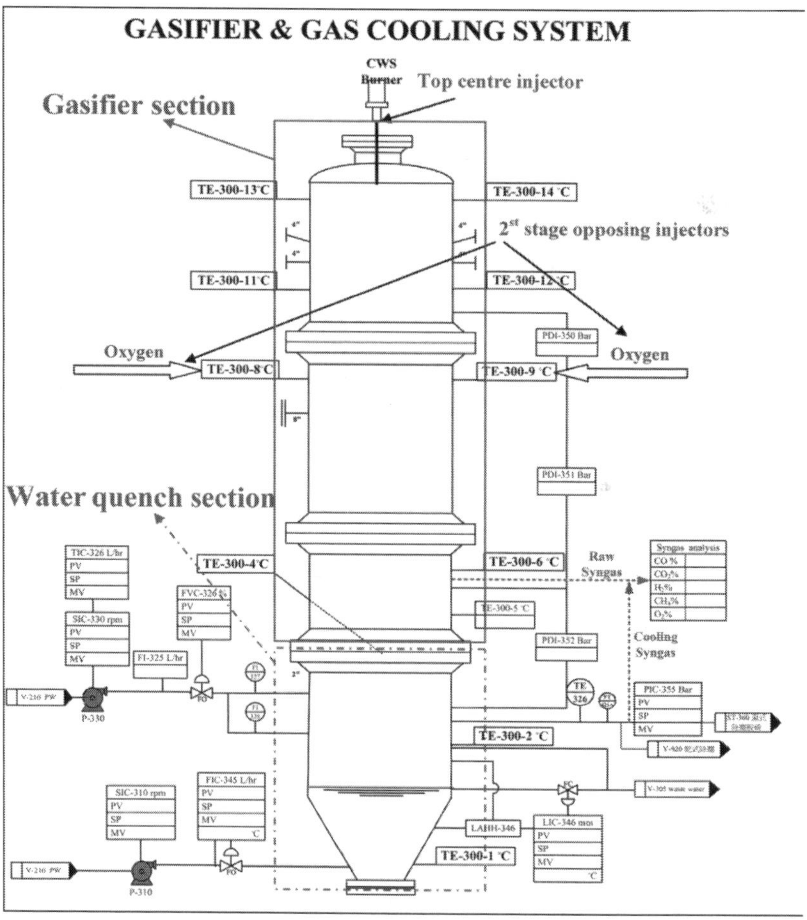

Figure 1. ITRI coal-slurry fed, oxygen blown, downdraft entrained-flow gasifier. The outline of the gasifier is based on the outside diameter of the refractory.

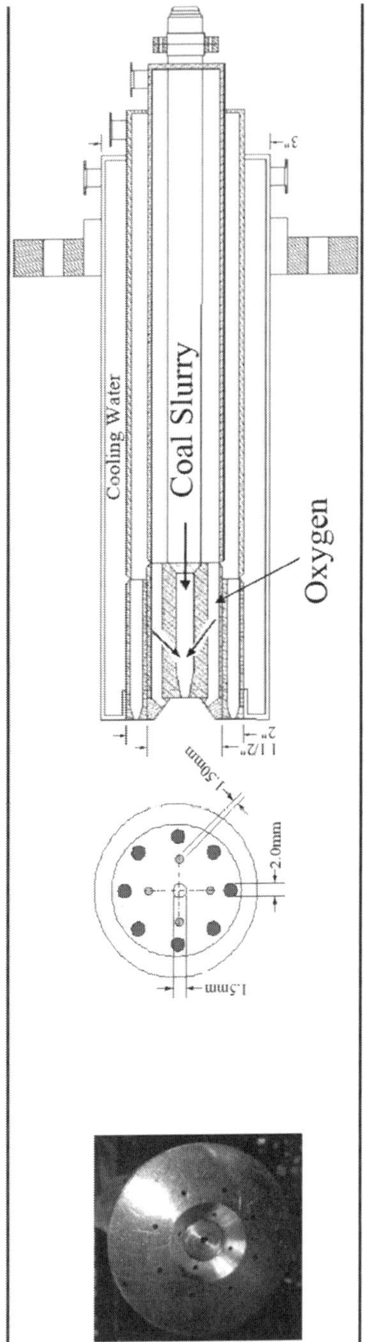

Figure 2. The internally premixed slurry-coal/oxygen injector with the coal-slurry fed in the center with oxygen fed through the co-centric outer annular pipe.

The gas temperatures are sampled with thermocouples being placed in cavities recessing one-inch in the wall. The gasification process occurs in a down-drafted mode with the syngas exiting near the bottom of the gasifier and continuously entering the quenching section. A detailed discussion about the quench section has been documented in Wang et al. [2] .

With the purpose of finding an optimum oxygen feeding ratio between injections from the top and a second location about 1/3 from the top of the gasifier, a series of experiments were conducted. The experiment is first conducted with 100% oxygen injected from the top, but it is discovered that the temperature is too high on the top and too cold in the middle section of the gasifier. Moreover, the carbon conversion ratio is poor, at around 72%. To improve the temperature distribution and carbon conversion ratio, part of the oxygen up to 60% (Figure 1), is diverted from the top injector to the side injectors in the middle section. This is done to hopefully provide more effective oxidation and hence more energy for feeding the gasification process in the middle section. The results are encouraging and show that the previous low temperatures in the middle and lower sections of the gasifier increased while the high temperature on the top reduced, resulting in a more uniform temperature distribution along the entire gasifier. More detailed analysis of the experimental results will be presented together with the computational results later.

3. COMPUTATIONAL MODEL

3.1. Global Gasification and Chemical Reactions

This study deals with the global chemical reactions of coal gasification [3] that can be generalized in reactions (R1.1) through (R1.10) in Table1

In this study, the methanation reactions are not considered since the production of methane is negligible under the studied operating conditions. The volatiles are modeled to go through a two-step thermal cracking process (R7) and gasification processes (R8, R9) with CH_4 or C_2H_2 as the intermediate products. The coal used in this study is Chinese Shanxi Pinghsuo coal, and its compositions are given in Table2 The compositions of volatiles are derived from the coal's heating value, proximate analysis, and ultimate analysis.

For the reaction rate of water-gas shift (WGS) reaction (R5), Lu and Wang [8] [9] tested Jones' rate by comparing the syngas composition results with the experimental data and, due to the difference in operating pressure and other boundary conditions, they purposely reduce the rate constant to A

= 2.75 to match the experimental data. Consequently, A = 2.75, E = 8.38 ′ 10^7 is used for the WGS reaction rate in this paper.

Table 1. Summary of reaction rate constants referenced in this study.

	Reactions	Reaction Type	Reaction heat, ΔH°_R (MJ/kmol)	$k = AT^n exp(-E/RT)$ (n = 0)		Reference
				A (s^{-1})	E (J/kmol)	
			Heterogeneous Reactions			
R 1	$C(s) + ½ O_2 \rightarrow CO$	Partial combustion	−110.5	0.052	6.1×10^7	Chen *et al.* [4]
R 2	$C(s) + CO_2 \rightarrow 2CO$	Gasification, Boudouard reaction	+172.0	0.0732	1.125×10^8	Chen *et al.* [4]
R 3	$C(s) + H_2O \rightarrow CO + H_2$	Gasification	+131.4	0.0782	1.15×10^8	Chen *et al.* [4]
			Homogeneous Reactions			
R 4	$CO + ½ O_2 \rightarrow CO_2$	Combustion	−283.1	2.2×10^{12}	1.67×10^8	Westbrook and Dryer [5]
R 5	$CO + H_2O(g) \leftrightarrow CO_2 + H_2$	Water Gas shift	−41.0	2.75×10^{10} #	8.38×10^7	Jones and Lindstedt [6]
R 6	$CO + 3H_2 \leftrightarrow CH_4 + H_2O$	Methanation	−205.7	$k_f = 4.4 \times 10^{11}$	1.68×10^8	
				$k_b = 5.12 \times 10^{-14}$	2.73×10^4	Benyon [7]
R 7	$CH_{2.997}O_{0.3456} \rightarrow 0.3456CO + 0.1897H_2 + 0.6544CH_4$ (Two-step Volatiles Cracking)		+12.088	Eddy dissipation		
R 8	$CH_4 + ½ O_2 \rightarrow CO + 2H_2$ Volatiles gasification via CH_4		−35.71	Eddy dissipation		N/A
R 9	$C_2H_2 + O_2 \rightarrow 2CO + H_2$ Volatiles gasification via C_2H_2		−447.83	Eddy dissipation		
R 10	$H_2 + ½ O_2 \rightarrow H_2O$	Oxidation	−242	6.8×10^{15}	1.68×10^8	Jones and Lindstedt [6]

1) All ΔH°_R at 298 K and 1 atm. 2) "+" Endothermic (absorbing heat), "−" Exothermic (releasing heat). # The value of A in [6] has been modified as 2.75 in this study.

Table 2. The proximate and ultimate analyses of Chinese Shanxi Pinghsuo coal.

Coal	Pingshuo Coal
Proximate Analysis, wt%	
Moisture	4.31
VM	32.60
Ash	13.52
Fixed Carbon	49.57
Higher Heating Value (HHV) (kcal/kg)	6612
Ultimate Analysis, wt%	
Moisture	2.04
Ash	12.69
C	70.94
H	4.78
N	0.78
S	0.58
O	8.19

3.2. Computational Methodology

The computational model and submodels (devolatilization, reactions, particle dynamics, gasification) used in the study are the same as developed by Silaen and Wang (2010) [10] , as well as Lu and Wang (2013) [8] [9] . Therefore, the governing and associated equations and detailed modeling intricacies are not repeated here, but they are briefly summarized below.

The time-averaged steady-state 3-D Navier-Stokes equations as well as the mass and energy conservation equations are solved. Species transport equations are solved for all gas species involved. It is well known that turbulence modeling has always been a sensitive issue for flow simulations because each turbulence model usually works only for some specific flow behaviors and none of any turbulence model can predict all the flows with high fidelity. Since investigation of the effect of turbulence models on gasification process is not an objective of this study, selection of an appropriate turbulence model is guided by a previous research performed by Silaen and Wang [10]. They applied five turbulence models (standard k-ε, k-ω, RSM, k-ω SST, and k-ε RNG) and reported that the RSM model predicted the best results. However, the standard k-ε turbulence model yields reasonable results without requiring significant computational time when compared to the more elegant RSM model. Therefore, the standard k-ε turbulence model is used in this study to provide closure. Enhanced wall function and variable material property are used. The P1 model is used for computing the radiation effect.

The flow (continuous phase) is solved in Eulerian form as a continuum while the particles (dispersed phase) are solved in Lagrangian form as a discrete phase. Stochastic tracking scheme is employed to model the effects of turbulence on the particles. The continuous phase and discrete phase are communicated through drag forces, lift forces, heat transfer, mass transfer, and species transfer. The finite-rate combustion model is used for the heterogeneous reactions, but both the finite-rate and eddy-dissipation models are used for the homogeneous reactions, and the smaller of the two is used as the reaction rate. The finite-rate model calculates the reaction rates based on the kinetics while the eddy-dissipation model calculates based on the turbulent mixing rate of the flow. Gasification or combustion of coal particles undergoes the following global processes: 1) evaporation of moisture; 2) devolatilization; 3) gasification to CO and 4) combustion of volatiles, CO, and char. The Chemical Percolation Devolatilization (CPD) model [11] -[13] is chosen as the devolatilization model based on the finding by Silaen and Wang (2010) [10] that the Kobayashi two-competing rates devolatilization model [14] is very slow while the CPD model gives a reasonable result.

For solid particles, the rate of depletion of the solid, due to a surface reaction, is expressed as a function of kinetic rate, solid species mass fraction on the surface, and particle surface area. The reaction rates are all global net rates, i.e., the backward reaction, calculated by equilibrium constants, are included in the global rate. Therefore, the finite rate employed in this study implicitly applies to the local equilibrium approach. Reaction rate constants used in this study are summarized in Table1

For liquid droplets, water evaporates from the particle's surface when the temperature is higher than the saturation temperature (based on local water vapor concentration). The evaporation rate is controlled by the water vapor diffusion from the droplet surface to the surroundings until 100% relative humidity is achieved. When the boiling temperature (determined by the air-water mixture pressure) is reached, water evaporation rate is controlled by the heat transfer rate at the droplet surface. After the moisture is evaporated, due to either high temperature or low moisture partial pressure, the vapor diffuses into the main flow and is transported away. Droplet breakup model and coalescence model have been applied for tracking the droplet movement. Please refer to Silaen and Wang [10] for details, which are not repeated here. Since this paper focuses on using an established existing CFD model and submodels to study the effect of distributed oxygen injection rather than develop a new model, no review of other research group's modeling efforts is conducted here.

3.3. Computational Domains

The computational domain and elements on the gasifier wall are shown in Figure 3. The computational domain contains roughly 1.1 million elements. A grid sensitivity study of 3-D geometries is conducted. Three different grids are used including a coarse grid (0.4 million cells), a medium grid (1.1 million cells), and a fine grid (1.6 million cells). The results of the temperature distribution along the gasifier show a difference of 7% between the coarse and medium grids and 2% between the medium and the fine grids. In order to save computational time, the medium grid of 1.1 million cells is chosen.

FLUENT 12.0.16 from ANSYS, Inc. is used as the CFD solver. The simulation is steady-state and uses the segregated solver, which employs an implicit pressure-correction scheme and decouples the momentum and energy equations. The SIMPLE algorithm is used to couple the pressure and velocity. Second order upwind scheme is selected for spatial discretization of the convective terms and species. For the finite rate model where the Eulerian-Lagrangian approach is used, the iterations are conducted alternatively between the continuous and the dispersed phases. Initially, two iterations in the continuous phase are conducted followed by one iteration in the discrete phase to avoid the flame from dying out. Once the flame is stably established, five iterations are performed in the continuous phase followed by one iteration in the dispersed phase. The drag, particle surface reaction, and mass transfer between the dispersed and the continuous phases are calculated. Based on the dispersed phase calculation results, the continuous phase is updated in the next iteration, and the process is repeated. Converged results are obtained when the residuals satisfy mass residual of 10^{-3}, energy residual of 10^{-5}, and momentum and turbulence kinetic energy residuals of 10^{-4}. These residuals are the summation of the imbalance in each cell, scaled by a representative for the flow rate.

3.4. Boundary and Inlet Conditions

The summary of the studied cases is listed in Table3 In Case 1, a two-stage configuration is used with oxygen distribution of 100% - 0% between the top and the second stages. The coal is fed as slurry. The total mass flow rate of the coal slurry is 92 kg/hr (0.026 kg/s), and the total volume flow rate of oxidant is 36.8 Nm^3/s (0.0146 kg/s). The top and the second injection nozzles are both 5 mm in diameter. To match the experimental conditions, the operating pressure is set as 3 bars for Case 1. From Case 2 to Case 8, the oxygen distribution is 100% - 0%, 90% - 10%, 80% - 20%, 70% - 30%, 60% - 40%, 50% - 50%, 40% - 60% between the top and the second injection stages respectively. Total mass flow rate of the coal slurry is 85 kg/hr (0.024 kg/s), and the total volume flow rate of oxidant is 38.6 Nm^3/s (0.0153 kg/s). The operating pressure is 2.5 bars for Cases 2 to 8.

The above eight cases from 1 - 8 are set up with the same operating conditions as the experiments. The following five cases from Cases 9 - 12 increase the operating pressure to 25 bars at which no experiments have been conducted in the current facility. From Case 9 to Case 12, the oxygen distribution between the top and the second injection stages is 100% - 0%, 80% - 20%, 60% - 40%, 40% - 60% respectively. Total mass flow rate of the coal slurry is 85 kg/hr (0.024 kg/s), and the total volume flow rate of the oxidant is 17.4 Nm^3/s (0.006912 kg/s), which is 30% of the stoichiometric amount for complete combustion.

The oxidant is considered to be a continuous flow, and the coal particles are considered to be the discrete phase. The discrete phase includes the fixed carbon and liquid water droplets from both the inherent moisture content of coal (4.31% wt) and the injected water content form the coal slurry (57.1% wt). Other components of the coal, such as N, H, S, O, and ash, are injected as gas together with the oxidant in the continuous flow to keep the mass flow the same as in the experiments. N is treated as N_2, H as H_2, and O as O_2. S and ash are not modeled, and their masses are lumped into N_2. Since studying the effect of coal particle sizes on gasification performance is not in the objectives of this study, the coal slurry size is uniformly given as 50 mm for the purpose of conveniently tracking the change of particle sizes. The effect of coal particle size on gasification process has been discussed in [5] and is not investigated in this study.

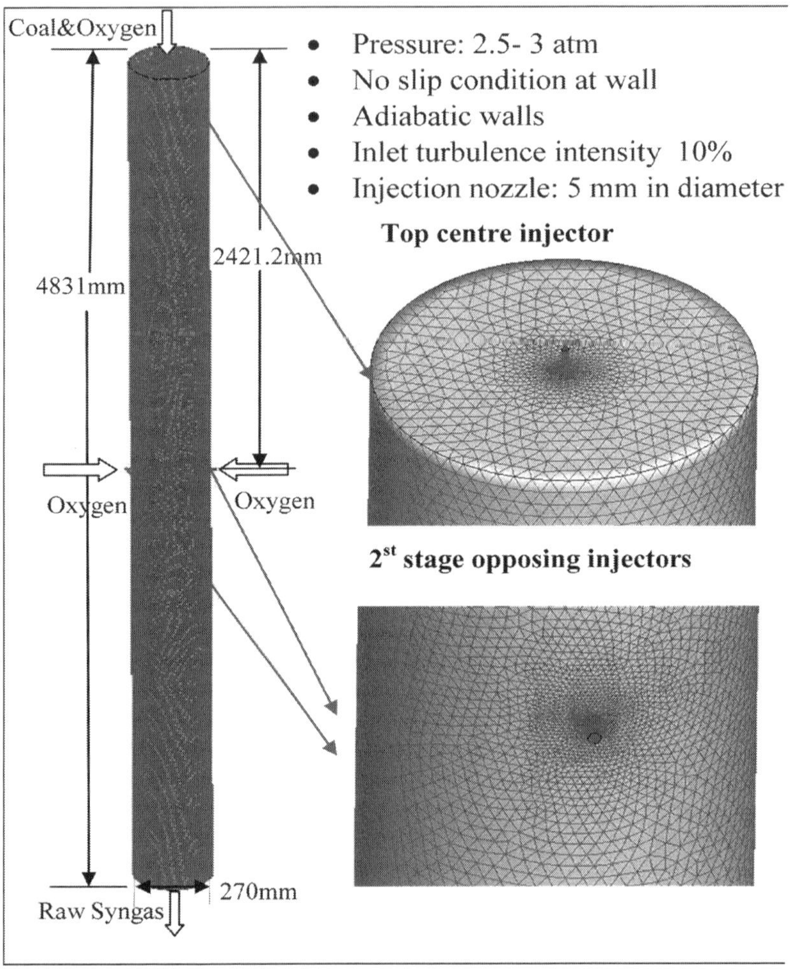

Figure 3. Meshed computational domain of the two-stage entrainedflow gasifier.

Table 3 . Parameters and operating conditions of the studied cases.

	Case 1	Cases 2 - 8	Cases 9 - 12 (CFD only)
O_2 distribution (top-second)	100% - 0%	(100% - 0%) - (40% - 60%)	(100% - 0%) - (40% - 60%)
Coal slurry injection (kg/s)	0.026	0.024	0.024
O_2 injection (kg/s)	0.0146	0.0153	0.009216
Operating pressure (bars)	3.0	2.5	25

The walls are assigned as adiabatic with internal emissivity of 0.8. In case of dry wall, interactions between the droplets and wall usually manifest three major phenomena, including reflect, break-up and trap. The actual phenomenon depends on the incoming Weber number of the droplet. Here, the Weber number is the ratio of kinetic energy of a droplet to the surface energy of a droplet ($We = \rho d_d v_d^2 / \sigma$). If the Weber number ($We_{in}$) is less than 10, the droplet will reflects elastically after it hits the wall. As the incoming We increases further to $We_{in} > 80$, the droplet falls into disintegration region which leads breaking-up the droplet to several small droplets. In the transition region of $30 < We < 80$, the droplet has the chance to either reflect or breakup. In the case of this study, since the Weber number is less than 10, the boundary condition of the discrete phase at walls is assigned as "reflect", which means the discrete phase elastically rebound off once reaching the wall. At the outlet, the discrete phase simply escapes/exits the computational domain. An area near the coal injection locations is initially patched with temperature of 1500 K to simulate the ignition process of a real operation. The limit of the highest temperature is assigned as 5000 K, and the limit of the lowest temperature is assigned as 400 K. This will remove potential runaway conditions caused by erratic, unreasonably high, or low temperatures during the iteration process.

4. RESULTS AND DISCUSSIONS

4.1. Comparison between CFD Results and Experimental Data

The computational cases 1 - 8 are conducted with the conditions matching those in the experiments. Among these eight cases, the detailed temperature and species distributions of Case 1 and Case 6 are exhibited to gain fundamental understanding of the thermal-flow behavior inside the gasifier. The conditions of these two cases are listed below.

- Case 1: Oxygen distribution (100% - 0%), coal slurry (92 kg/hr), oxidant (36.8 Nm^3/s or 0.0146 kg/s), and operating pressure (3 bars).
- Case 6: Oxygen distribution (60% - 40%), coal slurry (85 kg/hr), oxidant (38.6 Nm^3/s or 0.0153 kg/s), and operating pressure (2.5 bars).

4.1.1. Case 1 (100% - 0% for Oxygen Distribution)

Case 1 is one stage oxygen-blown operation with the oxygen distribution being 100% - 0% between the top and the second stages. Gas temperature and species mole fraction distribution on the horizontal and selected center vertical planes in the gasifier are shown in Figure 4. Since all the fuel is injected from the top, the gas temperature is higher in the top injection region than at the second stage location. Maximum gas temperature in the top injection stage reaches 2300 K. The dominant reaction in the top injection stage is the intense char combustion ($C + ½ O_2 \rightarrow CO$ and $CO + ½ O_2 \rightarrow CO_2$) as well as volatiles combustion via our 2-stage model in the first stage. Near the second stage, gasification reactions (mainly $C + ½ CO_2 \rightarrow CO$, $C + H_2O \rightarrow CO + H_2$) dominate. Due to the exothermic property of combustion process and endothermic property of the gasification reactions, the temperature drops gradually from top injection to the exit of the gasifier. Oxygen is completely depleted through the char combustion in the top injection region. The exit syngas composition calculated by CFD is shown in Table4

The mass-flow-rate weighted exit syngas temperature across over the entire exit section calculated by CFD is 1044 K and the near-wall average temperature is 986 K. They are 90 K and 32 K higher than the measured data, respectively. Note that the gas temperature was too hot to be measured by thermocouple, so the experimental temperature data was measured by thermocouples placed in cavities in the wall. Thus, it is not intended to compare the measured temperatures with the simulated core gas temperatures, rather the measured data are used to provide a reference for comparing the local near-wall syngas temperature at 5 mm from the wall.

Therefore, the comparison indicates that the average cross-sectional gas temperature is about 90 K hotter than the wall temperature and the near-wall temperature is about 32 K higher than the wall temperature.

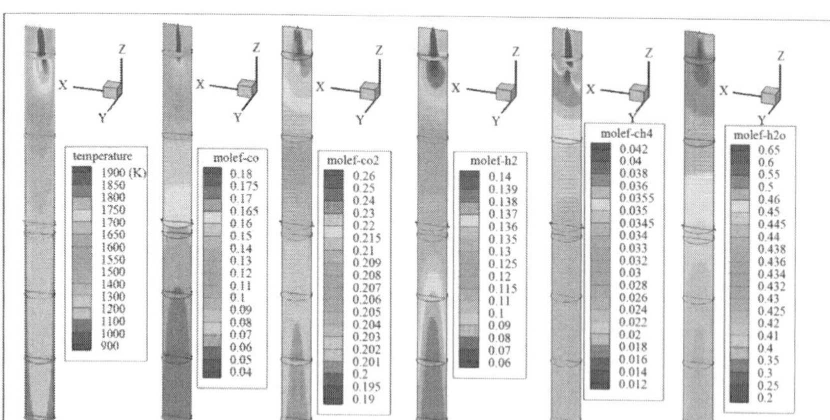

Figure 4. CFD simulated gas temperature and species mole fraction distributions for Case 1 (100% - 0% for oxygen distribution).

Table 4. Exit syngas composition of CFD results for Cases 1 and 6 and comparison of near-wall syngas temperature with measured temperatures in a wall cavity.

CFD syngas composition (Vol.)	Case 1	Case 6
CO	17%	18%
CO_2	20%	22%
H_2	14%	14%
H_2O	46%	44%
CH_4	3%	2%
CFD Temp (K)		
Cross-section average	1044	1550
Near-wall average	986	1172
Experi. Temp (K) in wall cavity	954	1150

4.1.2. Case 6 (60% - 40% for Oxygen Distribution)

Case 6 is the two-stage oxygen-blown operation with oxygen distribution of 60% - 40% between the first and the second stages. Gas temperature and species mole fraction distribution on the horizontal and center vertical planes in the gasifier are shown in Figure 5. The gas temperature is higher at the second stage injection region than at the top injection region. Maximum gas temperature reaches 2600 K at the second stage injection location and 1900 K at the top injection stage. The dominant reaction near the top injection location is the intense char combustion ($C + ½ O_2 \rightarrow CO$ and $CO + ½ O_2 \rightarrow CO_2$) as well as the volatiles oxidation via our two-stage model; then the endothermic gasification reactions occur in the region between the first and second stage injection locations as can be seen by increased CO and H_2 but reduced temperature. WGS also occurs, evident by reduced H_2O. At the second stage injection location, unburned char, CO and H_2quickly react with injected oxygen from two sides and release the second wave of combustion/oxidation heat ($C + ½ O_2 \rightarrow CO$, $CO + ½ O_2 \rightarrow CO_2$ and $H_2 + ½ O_2 \rightarrow H_2O$). The gasification reactions dominate the process after the oxygen is completely consumed in the region between the second stage injection location and the gasifier's exit.

Figure 6 and Figure 7 show CFD cross-sectional and mid-plane plots of the velocity vector, contours of temperature and mole fraction of CO_2 at the second injection stage for both Case 6 (60% - 40% for oxygen distribution) and Case 3 (90% - 10% for oxygen distribution) respectively. In Case 6, 40% oxygen, which is injected from two opposing injectors at the second stage, squeezes the main flow laterally to both sides perpendicular to the injection direction (top and bottom of Figure 6(a)and Figure 6(b)). The contours of temperature and mole fraction of CO_2 show that the chemical reactions are affected by the local flow behavior, which is demonstrated by the high temperature and CO_2 concentration presented towards top and bottom in Figure 6(a) and Figure 6(b). In Case 3, only 10% oxygen is injected at the second stage. The opposing injecting effect is not strong enough to block the main flow, so the combustion reaction only takes place below the second stage, which is shown by the contours of temperature and mole fraction of CO_2 in Figure 7.

The CFD calculated temperature is shown in Table 4 by two values. The first value (1550 K) is the mass-flow-rate weighted average temperature over the cross-section on the exit plane, and the second value (1172 K) is the mass-flow-rate weighted average temperature in the near-wall region (5 mm to the wall). The near-wall temperature is comparable to the experimental data (1150 K) within 22 K. Since the thermocouples were placed in cavities recessing in the wall, the measured data adequately reflects the relatively cooler syngas temperature near the wall. Figure 8 shows the temperature contour on the cross-section of the exit plane. It clearly shows that the temperature changes significantly from the center at around 2000 K to around 1150 K in the near-wall region.

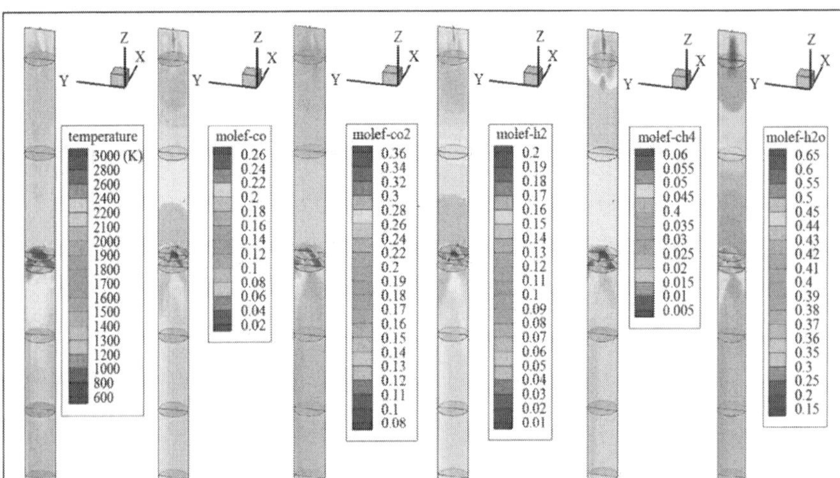

Figure 5. CFD simulated gas temperature and species mole fraction distributions for Case 6 (60% - 40% for oxygen distribution).

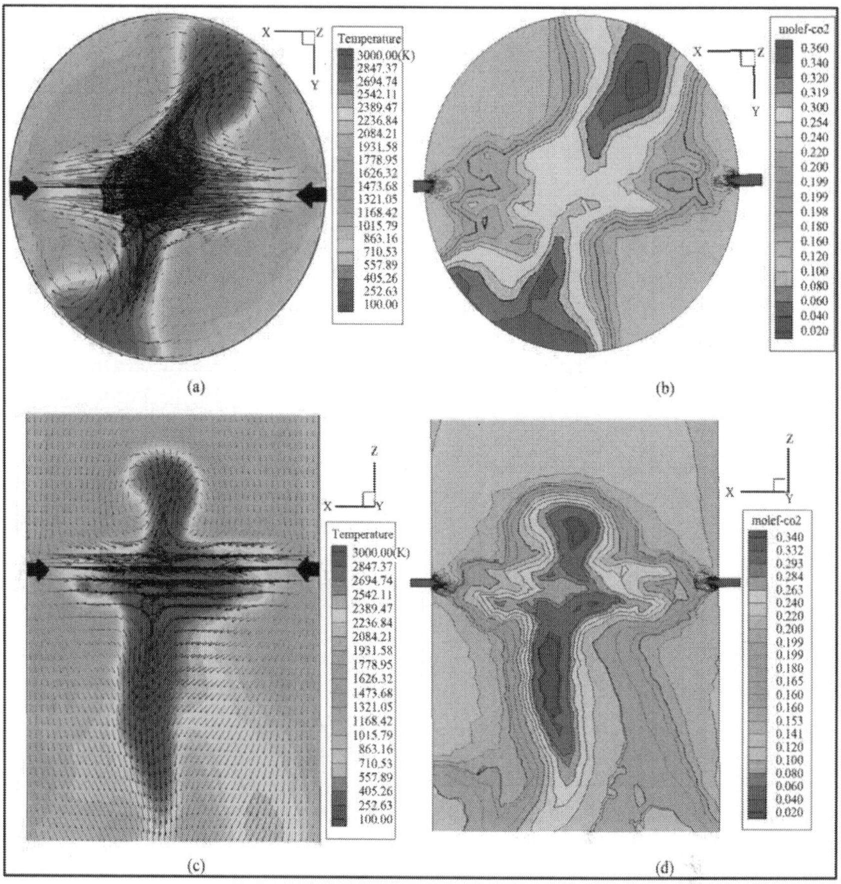

Figure 6. CFD cross-sectional and mid-plane plots of velocity vector, contours of temperature and mole fraction of CO_2 for Case 6 (60% - 40% for oxygen distribution). (a) Cross-sectional temperature contour and velocity vector at the second injection. (b) Cross-sectional mole fraction of CO_2 at the second injection. (c) Mid-plane temperature contour and velocity vector near second injection. (d) Mid-plane mole fraction of CO_2near second injection.

4.2. Effect of Oxygen Injection Distribution under the Experimental Condition (Operating Pressure: 2.5 Bars, O_2: 0.6 of Stoichiometric Amount)

To investigate the effect of different oxygen-distribution feeding schemes on the gasification performance under the low operating pressure condition, seven computational simulations cases corresponding to the experimental conditions (operating pressure: 2.5 bars, O_2: 0.6 of stoichiometric amount for complete combustion) are conducted with the oxygen fed from top at 100%, 90%, 80%, 70%, 60%, 50%, and 40% respectively. These cases are numbered from Case 2 to Case 8 as shown in Table3 The CFD results of these 7 cases are shown in Table 5, including the syngas composition at the exit, the syngas temperature at four locations (T_{exit} = TE-300-4, T2 = TE-300-8, T3 = TE-300-12, T4 = TE-300-13), carbon conversion ratio (CCR), and the higher heating value (HHV). The comparisons between CFD results and experimental data are shown graphically in Figure 9. The variations of simulated syngas compositions for different cases are shown in Figure 10.

Figure 9 shows that the experimental data and the CFD results yield the same trend of change for temperature distribution at the exit as the oxygen feeding distribution changes between two stages. When the percentage of oxygen injected from top reduces from 100% to 40%, the temperature decreases at the first stage (T1 and T3) and increases at the second stage (T2 and T_{exit}). This is because if less oxygen is injected from the top, then less char will go through the exothermic combustion process and less combustion heat is released. At the first glance, the variation trends appear to reverse around 60% - 50%. Therefore, the analysis and explanations are separated into two parts based on the oxygen distribution on the top: 100% - 60% vs. 60% - 40%. The CFD simulated syngas compositions are used to provide explanation of the variations of near-wall temperature distributions.

100% - 60% top O_2 injection-Figure 10 shows that CFD simulated mole fractions of CO and H_2 at the exit increase when the top injected oxygen reduces from 100% to 60%, but they reduce when the top injected oxygen decreases from 60% to 40%. The mole fraction of CO_2 has an opposite changing trend.

60% - 40% top O_2 injection-If the mass fraction of top injected oxygen continues to reduce below 60%, there is more unburned char going though the gasification process (mainly $C + ½ CO_2 \rightarrow CO$, $C + H_2O \rightarrow CO + H_2$), resulting in more production of CO and H_2 at the first stage. At the second injection stage, some of the injected oxygen reacts with CH_4 to produce CO and H_2. Therefore, the total amount of CO and H_2 increases and CO_2 decreases at the exit of gasifier. However, if more than 40% - 50% of the total oxygen is injected from the second injection location, more CO and H_2, which were produced through the gasification process at the first stage, react with the oxygen through the combustion process at the second stage and result in more CO_2 and H_2O. Therefore, as indicated in Figure 10, CO and H_2 decrease and CO_2 and H_2O increase at the exit of the gasifier when the portion of oxygen injected from the top is reduced from 60% to 40% based on the CFD results Meanwhile, the carbon conversion ratio increases, and the exit mole fraction of CH_4 reduces when the percentage of oxygen injected from the top reduces from 100% to 40%. The reason for this trend is that the unburned char and CH_4 react with oxygen through $C + ½ O_2 \rightarrow CO$ and $CH_4 + ½ O_2 \rightarrow CO + 2H_2$ at the second injection stage. When oxygen injected from the second stage increases beyond 40%, the carbon conversion ratio reaches 99%.

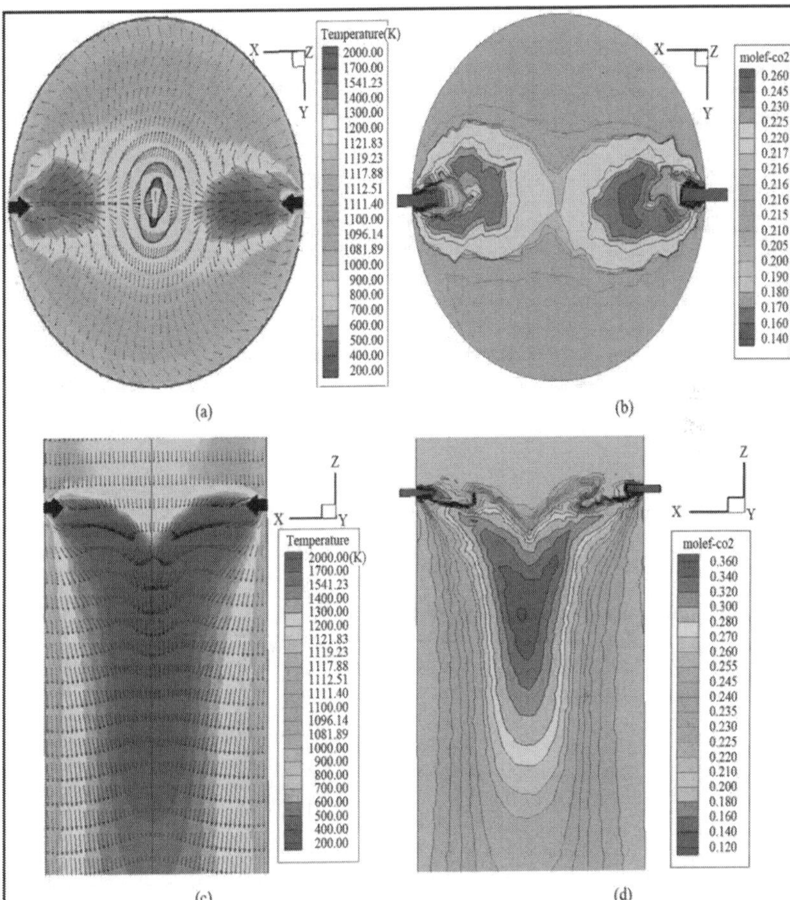

Figure 7. Cross-sectional and mid-plane plots of velocity vector, contours of temperature and mole fraction of CO_2 at second injection for Case 3 (90% - 10% for oxygen distribution). (a) Cross-sectional temperature contour and velocity vector at the second injection. (b) Cross-sectional mole fraction of CO_2 at the second injection. (c) mid-Plane temperature contour and velocity vector second injection. (d) Mole fraction of CO_2 at the second injection.

One of the advantages of employing the two-stage oxygen control is to reduce the wall temperature gradient, and hence, the thermal stress along the gasifier can be reduced. The experimental results in Figure 9(a) show that the overall near-wall temperature difference in the gasifier is reduced to a narrower range of 900°C - 1200°C (with a 60% - 40% two-stage oxygen injection split) from a wider range of 600°C - 1500°C (with 100% top injection). In contrast, the CFD result in Figure 9(b) shows that the case of 80% - 20% yields the most uniform syngas temperature within a range from 955°C to 1354°C, while the case of 60% - 40% in Table 5 yields the highest syngas HHV (127,713 kJ/kmole). The case of 80% - 20% yields the lowest peak temperature and the most uniform syngas temperature within a range from 810°C to 1150°C. No experimental data are available for comparing the 80% - 20% CFD case.

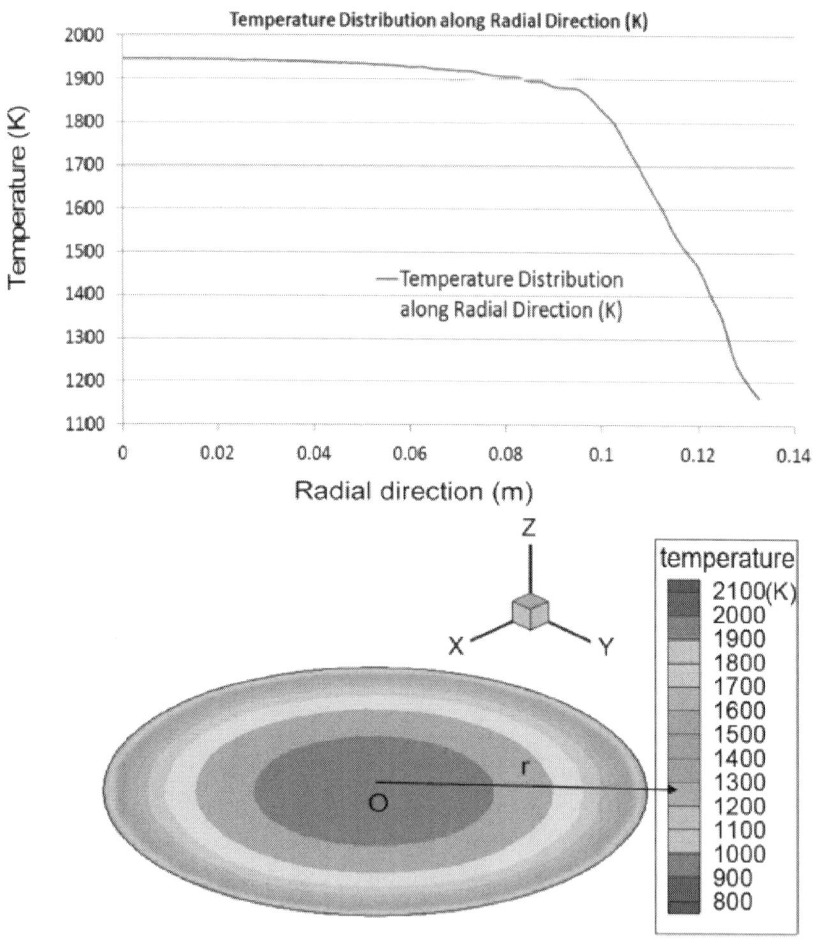

Figure 8. Gas temperature distribution at exit cross-sectional plane of the gasifier for Case 6 (60% - 40% oxygen distribution).

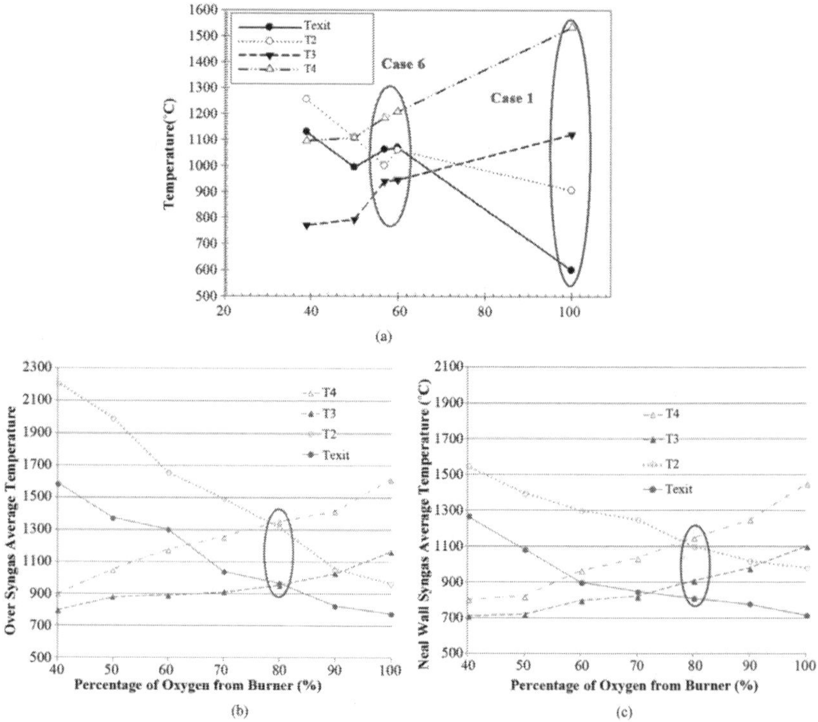

Figure 9. The effect of two-stage oxygen injection distribution on the gasifier wall temperature (a) experimental data (in wall cavities); (b) CFD results of mass-flow weighted syngas exit temperature; (c) CFD results for near-wall temperature (5 mm to the wall) for Cases 2 to 8 (Note: T_{exit}, T2, T3, and T4 in this figure correspond to TE-300-4, 8, 12, and 13 in Figure 1).

4.3. Effect of Oxygen Injection Distribution under High Operating Pressure Condition (25 Bars with 40% of Stoichiometric O_2)

Since most of the gasifiers are operated under the condition with high pressures above 25 bars and low sub-stochiometric oxygen feeding below 40%, the rest of the analysis will be focused on presenting the CFD simulation at a high pressure (25 bars) and low oxygen stoichiometric ratio (0.4). In total, four cases (Cases 9 to 12 as listed in Table 3) have been conducted.

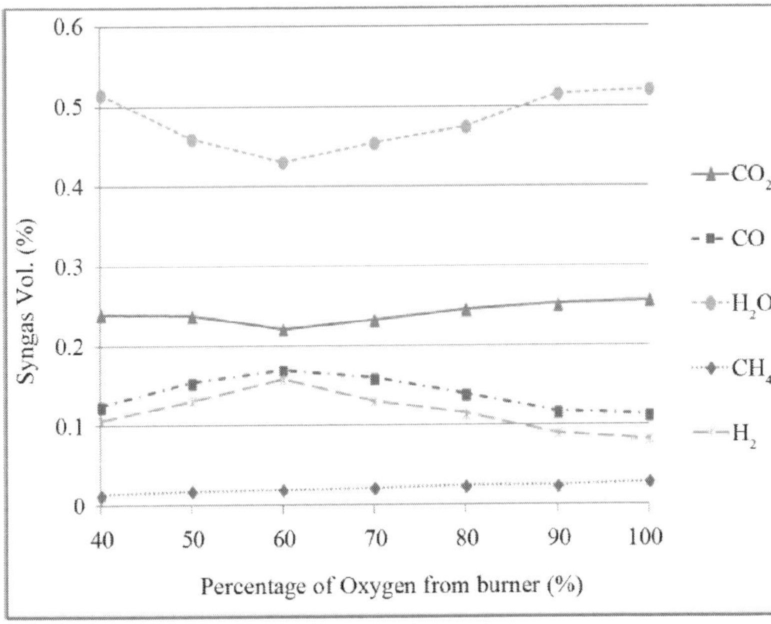

Figure 10. CFD simulated effect of two-stage oxygen injection distribution on the syngas composition for Cases 2 to 8.

Table 5. CFD results of syngas compositions, syngas temperatures at different locations, carbon conversion ratio, and HHV for Cases 2 to 8. $T_{exit, plane}$ is the mass flow weighted average over the cross-section of the exit plane. $T_{exit, near wall}$ is the circumferential average temperature at 5 mm from the wall.

O_2 distribution	Case 2 100% - 0%	Case 3 90% - 10%	Case 4 80% - 20%	Case 5 70% - 30%	Case 6 60% - 40%	Case 7 50% - 50%	Case 8 40% - 60%
CO_2 (Vol)	0.256	0.252	0.245	0.233	0.222	0.238	0.240
CO (Vol)	0.113	0.117	0.140	0.160	0.170	0.154	0.125
H_2O (Vol)	0.520	0.516	0.475	0.455	0.430	0.460	0.515
H_2 (Vol)	0.082	0.090	0.115	0.130	0.158	0.130	0.107
CH_4 (Vol)	0.029	0.025	0.025	0.022	0.020	0.018	0.013
T4 (K)	1892	1689	1627	1530	1450	1325	1174
T3 (K)	1439	1301	1228	1186	1164	1154	1071
T2 (K)	1237	1329	1590	1769	1934	2267	2484
$T_{exit plane}$ (K)	1052	1101	1242	1314	1550	1647	1859
$T_{exit, near wall}$ (K)	992	1052	1084	1124	1172	1356	1542
CCR	72.90%	74.62%	85.86%	93.02	98.88%	99.22%	99.94%
HHV (kJ/kmol)	73.180	78,455	97.559	110,776	127.713	109.078	89,284

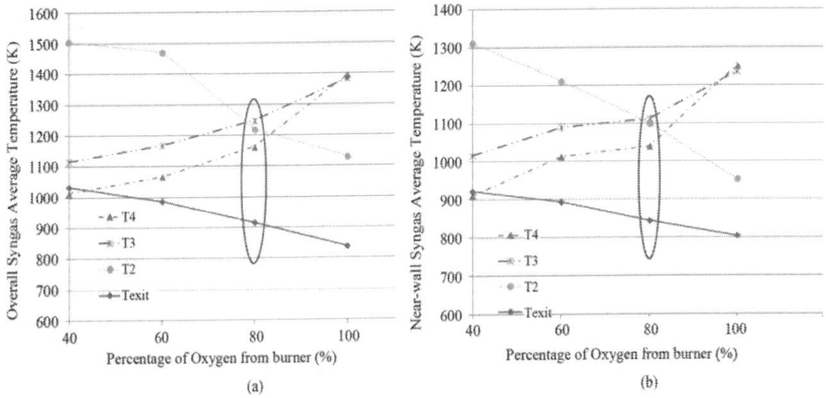

Figure 11. CFD simulated effect of two-stage oxygen injection distribution on the mass flow weighted syngas temperature at high pressure (25 bars) for Cases 9 to 12. (Note: The temperature locations T_{exit}, T2, T3, T4 in this figure correspond to TE-300-4, 8, 12, and 13 in Figure 1). (a) Average cross-sectional temperature. (b) Average near-wall temperature.

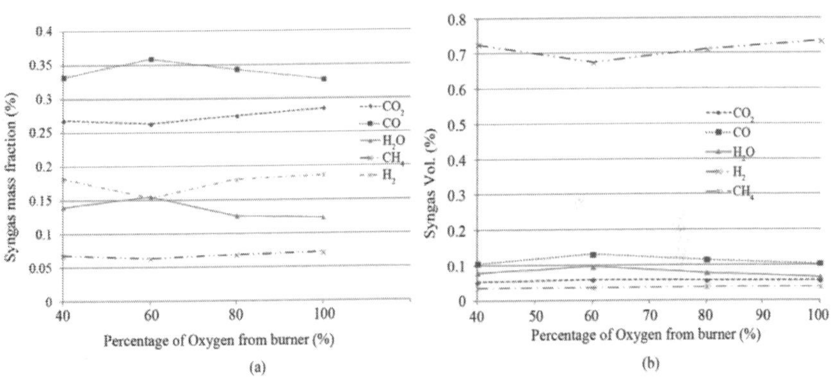

Figure 12. CFD simulated effect of two-stage oxygen injection distribution on the syngas composition at high pressure (25 bars) for Cases 9 to 12. (a) Mass fraction. (b) Volume fraction.

Simila to the discussion in section 4.2, the syngas composition and temperature at different locations, CCR and HHV for Cases 9 - 12 are provided in Table 6, Figure 11 and Figure 12respectively. The results are summarized below.

1) Compared to the results of the low operating pressure cases, the carbon conversion rates for different oxygen injection distribution are all above 95% in high operating pressure cases. Two mechanisms are considered to contribute to this higher carbon conversion rate: a) The partial pressure of

the CO_2, H_2O, and even O_2 increases by a factor of 10. The char reaction rate expression contains the partial pressure of these gases at the coal particle surface, so the char conversion rate (mass per time) increases notably as the operating pressure increases. b) The longer particle residence time could also contribute to achieving a higher carbon conversion ratio. The average residence time is about 6 - 7 seconds for the high pressure cases while it is 3 - 4 seconds for the low-pressure cases.

2) Both the high operating pressure and low pressure cases follow the same trend of change of exit average temperature as the oxygen feeding distribution changes between two stages.

Table 6. CFD predicted syngas conditions at 25 bars for Cases 9 to 12.

O_2 distribution	Case 9 100% - 0%	Case 10 80% - 20%	Case 11 60% - 40%	Case 12 40% - 60%
CO_2 Vol (Mass)	0.056 (0.285)	0.057 (0.275)	0.059 (0.263)	0.053 (0.268)
CO Vol (Mass)	0.103 (0.328)	0.115 (0.343)	0.131 (0.359)	0.105 (0.332)
H_2O Vol (Mass)	0.067 (0.123)	0.078 (0.126)	0.097 (0.155)	0.078 (0.140)
H_2 Vol (Mass)	0.734 (0.186)	0.712 (0.180)	0.674 (0.153)	0.726 (0.182)
CH_4 Vol (Mass)	0.037 (0.072)	0.037 (0.069)	0.036 (0.063)	0.035 (0.068)
T4 (K)	1395	1162	1067	1014
T3 (K)	1383	1247	1169	1116
T2 (K)	1129	1220	1472	1506
$T_{exit\ Plane}$ (K)	838	915	986	1033
$T_{exit,\ near\ wall}$ (K)	804	845	894	922
CCR	99.99%	95.34%	98.88%	99.97%
HHV (kJ/kmol)	392,295	391,271	376,653	395,495

3) Consistent with the low pressure cases, the changing trend of syngas composition reverses at the 60% - 40% case. The 80% - 20% case produces the lowest peak temperature and most uniform temperature in the gasifier, but with the lowest CCR.

4) As more oxygen is injected at the second stage, CO and CO_2 have the same trend of change as in the low pressure condition; however, H_2O and H_2 change in the opposite direction.

5) The major differences between the high and the low pressure cases are: a) the average temperature is much lower at each corresponding location in the 25 bars condition; b) H_2dominates the syngas composition (about 70% vol.) at 25 bars case.

6) The 40% - 60% case produces the syngas with the highest HHV in the 25 bar case vs. 60% - 40% case in the 2.5 bars cases, but with the highest peak wall temperature.

5. CONCLUSIONS

1) The experiments at 2.5 bars show the 60% - 40% oxygen distribution yields the most uniform wall temperature along the gasifier.

2) In cases of split two-stage oxygen injections from 100% to 40% with 0.6 stoichiometric oxygen at 2.5 bars, the CFD predicts a) the similar trend of changing temperature as in the experimental data; b) the 80% - 20% case yields the lowest peak temperature and the most uniform gas temperature distribution along the gasifier; and c) the 60% - 40% case produces the syngas with the highest HHV.

3) The CFD's prediction of the effect of oxygen distribution under 25 bars operating pressure case shows that a) the carbon conversion ratio for different oxygen injection schemes are all above 95%; b) H_2 (about 70% vol.) dominates the syngas composition at exit; c) same as in the 2.5 bar cases, the 80% - 20% case yields the lowest peak temperature and most uniform gas temperature distribution along the gasifier; and d) the 40% - 60% case produces the syngas with the highest HHV.

4) Chemical reactions are affected by the local flow behavior. Strong opposing injection of oxygen at second stage squeezes the main flow laterally on the vertical planes perpendicular to the injection direction, making combustion spread accordingly in the vertically plane at the second stage.

Both experimental data and CFD predictions verify the hypothesis that it is feasible to reduce the peak temperature and achieve more uniform temperature in the gasifier by adequately controlling two-stage oxygen injection with only minor changes of CCR and HHV under the high-pressure (25 bars) condition. However, the variations of CCR and HHV are notable under the low pressure (2.5 bars) condition.

After this study was completed, we realized that Tsinghua University in China has already successfully developed an oxygen-staged gasifier [15] . One of the merits claimed by the Tsinghua research group was that the oxygen-staged scheme helped to reduce the syngas temperature and thus significantly extends the fuel injector's life expectancy. It appears that the success of Tsinghua's oxygen-stage scheme supports the hypothesis of this study. Since no detailed technical information has been released by Tsinghua, the information generated by this study from both experiments and CFD simulation are considered to be useful reference to the gasification community.

ACKNOWLEDGEMENTS

This study was partially supported by the Louisiana Governor's Energy Initiative via the Clean Power and Energy Research Consortium (CPERC) administered by the Louisiana Board of Regents. The ITRI gasifier was funded by the Taiwan Bureau of Energy.

REFERENCES

1 Wang, T., Silaen, A., Hsu, H.W. and Shen, C.H. (2010) Investigation of Heat Transfer and Gasification of Two Different Fuel Injectors in an Entrained-Flow Gasificr. ASME Journal of Thermal Science and Engineering Applications, 2, 011001/1-10.

2 Wang, T., Lu, X., Hsu, H.W. and Shen, C.H. (2011) Investigation of the Performance of Syngas Quench Cooling Design in a Downdraft Entrained Flow Gasifier. Proceedings of the 28th International Pittsburgh Coal Conference, Pittsburgh, 12-15 September 2011, 1-18.

3 Smoot, L.D. and Smith, PJ. (1985) Coal Combustion and Gasification. Plenum Press, New York.

4 Chen, C., Horio, M. and Kojima, T. (2000) Numerical Simulation of Entrained Flow Coal Gasifiers. Chemical Engineering Science, 55, 3861-3833.

5 Westbrook, C.K. and Dryer, F.L. (1981) Simplified Reaction Mechanisms for the Oxidation of Hydrocarbon Fuels in Flames. Combustion Science and Technology, 27, 31-43.

6 Jones, W.P. and Lindstedt, R.P. (1988) Global Reaction Schemes for Hydrocarbon Combustion. Combustion and Flame, 73, 233.

7 Benyon P. (2002) Computational Modelling of Entrained Flow Slagging Gasifiers. PhD Thesis, School of Aerospace, Mechanical & Mechatronic Engineering. University of Sydney, Sydney.

8 Lu, X. and Wang, T. (2013) Water-Gas Shift Modeling in Coal Gasification in an Entrained-Flow Gasifier Part 1: Development of Methodology and Model Calibration. Fuel, 108, 629-638.

9 Lu, X. and Wang, T. (2013) Water-Gas Shift Modeling in Coal Gasification in an Entrained-Flow Gasifier Part 2: Gasification Application. Fuel, 108, 620-628.

10 Silaen, A. and Wang, T. (2010) Effect of Turbulence and Devolatilization Models on Gasification Simulation. International Journal of Heat and Mass Transfer, 53, 2074-2091.

11 Fletcher, T.H., Kerstein, A.R., Pugmire, R.J. and Grant, D.M. (1990) Chemical Percolation Model for Devolatilization: 2. Temperature and Heating Rate Effects on Product Yields. Energy and Fuels, 4, 54-60.

12 Fletcher, T.H. and Kerstein, A.R. (1992) Chemical Percolation Model for Devolatilization: 3. Direct Use of 13C NMR Date to Predict Effects of Coal Type. Energy and Fuels, 6, 414-431.

13 Grant, D.M., Pugmire, R.J., Fletcher, T.H. and Kerstein, A.R. (1989) Chemical Percolation of Coal Devolatilization Using Percolation Lattice Statistics. Energy and Fuels, 3, 175-186.

14 Kobayashi, H., Howard, J.B. and Sarofim, A.F. (1976) Coal Devolatilization at High Temperatures. Symposium (International) on Combustion, 16, 411-425.

15 Zhang, J. (2013) Innovativeness and Development of Tsinghua Gasification Technology. Gasification Technology Conference, Colorado Spring, 13-16 October 2013, 17 p.

CHAPTER 10

Sustainable Energy Portfolios for Small Island States

Sándor Szabó*, Ioannis Kougias, Magda Moner-Girona and Katalin Bódis

European Commission, Joint Research Centre, Institute for Energy and Transport, Renewables and Energy Efficiency Unit, Via E. Fermi, 2749 Ispra, Italy

ABSTRACT

The study presents a cost effective electricity generation portfolio for six island states for a 20-year period (2015–2035). The underlying concept investigates whether adding sizeable power capacities of renewable energy sources (RES) options could decrease the overall costs and contribute to a more sustainable, indigenous electricity generation at the same time. Often, island states rely on fossil fuels which, apart from dependence on foreign resources, also includes an additional, significant transport cost. This is an extra motive to study the extent in which island states represent primary locations for RES technologies. For the aims of the present study an optimization model has been developed and following numerous runs the obtained results show that installing PV and battery capacities can delay-reduce the huge investments in fossil options in early periods. Thus, investment on RES can have a positive, long-term effect on the overall energy mix. This prompt development can happen without adding new subsidies but there is a need to address the existing socio-economic barriers with intelligent design of financing and economic instruments and capacity building as discussed in the conclusions.

KEYWORDS

renewable energy; sustainable development; solar; PV; battery; fossil fuels; optimization; harmony search algorithm; island energy portfolio

1. INTRODUCTION

The existing international cooperation in the energy domain concentrates its efforts on accelerating access to clean energy by promoting renewable energy sources (RES), energy efficiency and supporting schemes for investment in energy infrastructure and sustainable energy technologies. The present paper focuses on the third measure: applying sustainable energy technologies in an island state setting. The United Nations (UN) identifies sustainable development as a simultaneous enhancement of three interconnected goals: economic development, social development and environmental protection [1].

Island states are generally more vulnerable than continental regions in all three aspects due to the adverse effects of global climate change. Many of island states are low-lying coastal countries and have dominant part of their population living on these coasts with limited means to adapt to changes in climate conditions. The predicted climate change effects and sea-level rise, as projected by the Intergovernmental Panel on Climate Change (IPCC) [2] in the business-as-usual scenario, implies more severe consequences on the island states than the comparable mainland countries in terms of human, social and economic well-being.

As far as socio-economic development is concerned, the lack of adaptation capabilities is not the only and most imminent exposure of these islands to the risks caused by global economic trends. The universal development tendency that per capita electricity consumption progresses proportionally with the GDP, applies likewise to island states. Data in consistent time series is not always available for island states. Therefore, the present study analyzed only countries where at least basic energy-related data could be derived from the often fragmented and inconsistent information for islands. Figure 1 andFigure 4 illustrate processed information for 22 small island states, that has been collected from three different sources: [3,4,5]. The selected 22 islands are currently at different development stages. Thus, any analysis of the GDP-consumption on an individual-island basis, would produce non-comparable results. For this reason Figure 1 depicts a snapshot of the current GDP-consumption status of the selected island states, where the above mentioned tendency is clearly traceable. And since most of the islands rely heavily on fossil fuel sources, their dependence on foreign energy resources would increase with the GDP growth unless this underlying tendency changes.

The major driver behind the present study is the opportunities created by the unprecedented price decrease in the RES technologies in the last 15 years. Energy technologies considered unaffordable, have experienced so drastic price decrease that can become the prime choice in sustainable energy development. However, pipeline projects in island countries do not reflect this shift: the electricity master plans for the islands still dominantly involve fuel generation units. Amongst RES technologies PV has become an option applicable on islands with continuously decreasing price (Figure 2). The other RES option, wind turbines, have also showed impressive price development, but their price have stabilized at a higher level (Figure 3). Siting wind power also requires more localized information ([4]). Therefore, the present study focuses on modeling the potential role of PV technology. In this analysis we could rely on an internationally recognized continental level resource mapping tool (PVGIS) and a similar methodology that was used for rural electrification in Africa [6,7].

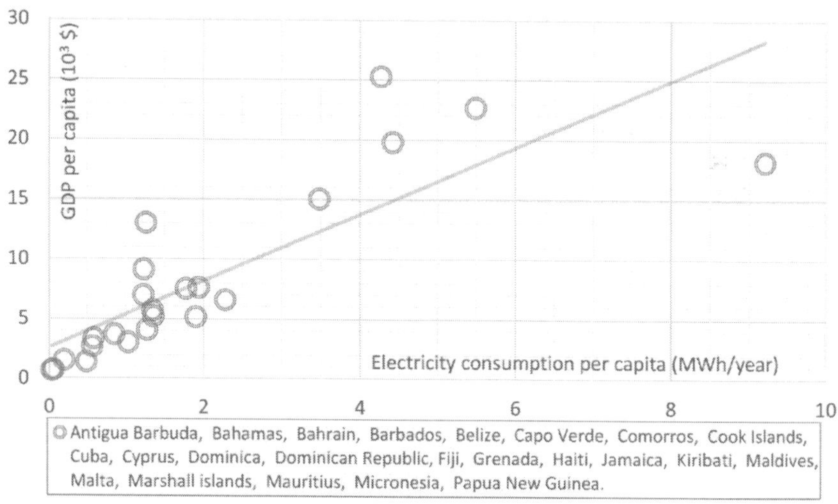

Figure 1. Electricity consumption and GDP per capita for the studied island states.

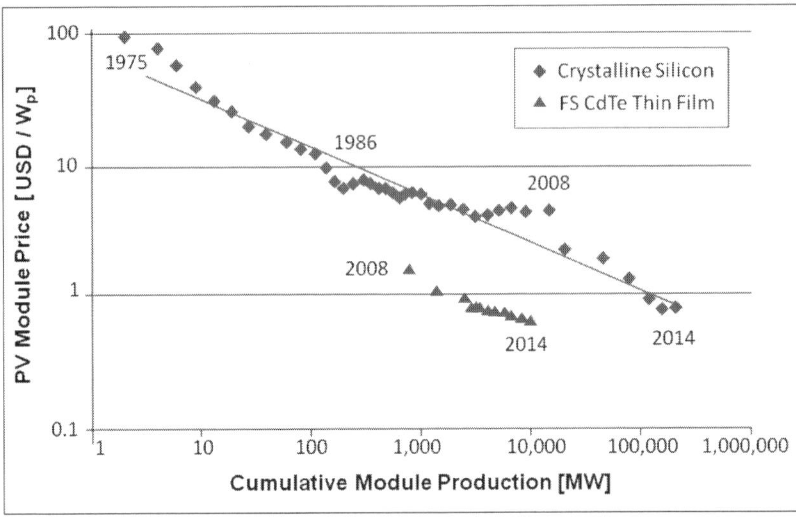

Figure 2. Global PV module price learning curve (1975–2014) [8].

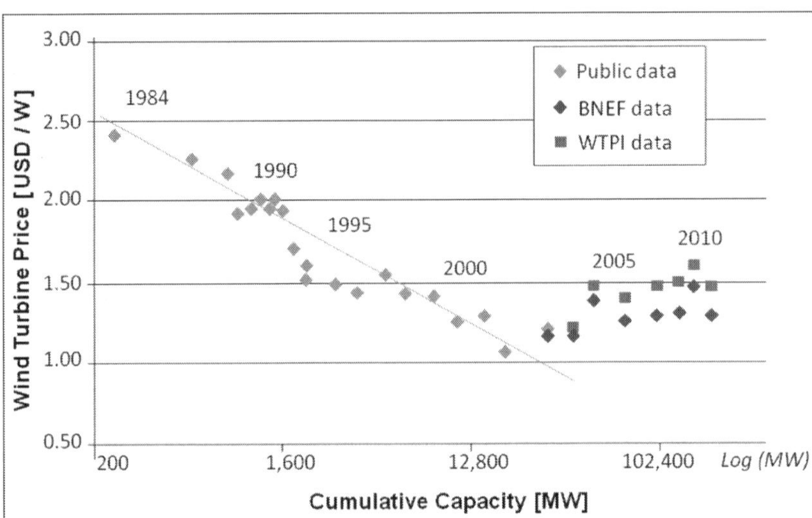

Figure 3. Global wind turbine price learning curve (1984–2010) [9].

1.1. Current Energy Status of Island States

Various Greek islands can be referred as a profound example on how the subsidies on heavy oil electricity production could lead to an unsustainable energy mix. In many cases, the annual cost of fuel shipped to the Greek islands exceeds €2,250/inhabitant [10]. However, consumers in these islands do not pay the full costs, as the Greek government covers most of the expenses from taxation. Independent island states do not have this privilege of "energy bills' dilution". Besides, even if they could use such mechanisms, they would not afford the additional cost from their own budgets.

The underlying drive behind the present analysis is that while many island states announce ambitious sustainable energy targets and pathways, the available data sources on the pipeline projects still show the dominance of the fossil fuel based technologies [3,4]. Figure 1 indicates the correlation between per capita electricity consumption and the corresponding GDP levels in the islands.

The aim of the present analysis is to investigate how this worldwide tendency relates to economic sustainability. There is a wide choice of technologies that can meet the increasing electricity needs, ranging from high environmental footprint technologies to entirely sustainable solutions.

In Figure 4 it appears that as prosperity grows the islands tend to rely more on imported fossil fuels for electricity production (mostly heavy oil) instead of relying on renewable energy sources to meet their energy needs. The size of the bubbles indicates population.

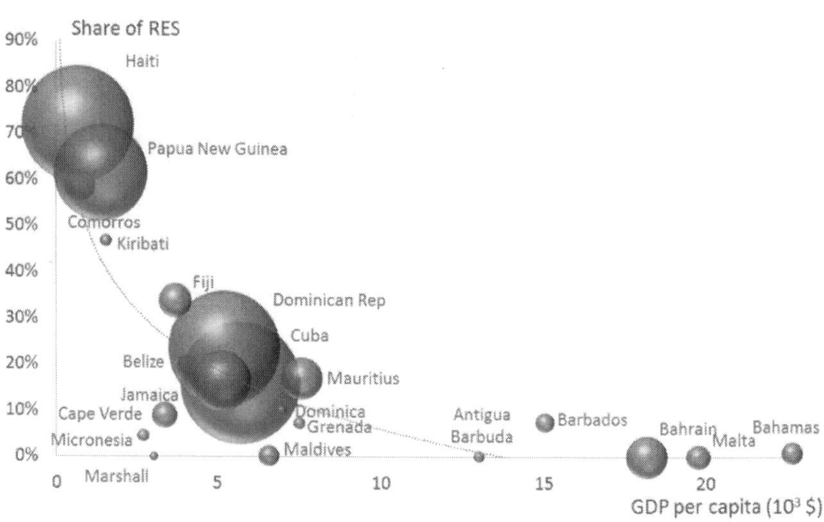

Figure 4. Share of RES and GDP per capita at the selected island states.

However, evidence is growing in the literature [11,12] that distributed generation options represent cost optimal solutions for many rural areas in

Sub-Saharan Africa as the grid extension options represent more expensive electricity provision for places far away from the existing power grids [6,7]. This could especially appear to be the case for the island states which are even more distant from fuel supply sources than most of the rural African communities.

The central question of the sustainable energy provision on islands is whether the least-cost electricity for the consumers can be delivered by the currently dominant fossil fuel electricity generation technologies or by cost-effective renewable energy solutions. The economic vulnerability and dependence of these states are shown by the fact that while electricity production moves globally away from oil for environmental protection and economic reasons, in most cases islands' production still depends mainly on it. EIA [13] and IEA [14] statistics indicate that, on a global scale, the share of oil is 5% (2012), decreasing from the previous 25% in 1973 (Figure 5). Contrary to that, oil electricity production represents >90% in Cyprus, 99% in Channel island and 81% in Malta. These countries have not managed to substitute oil with a lower-cost or lower-emission source during the last four decades and are exposed to the price fluctuation of a resource that is imported from distant sources.

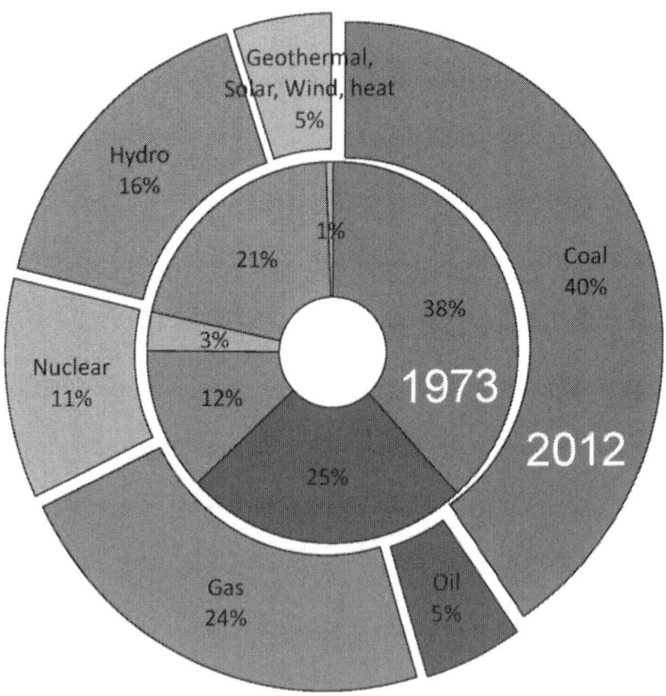

Figure 5. Electricity generation share on a global scale (1973 *vs.* 2012).

In [15] the authors have made an optimisation exercise for the European countries to see the optimal level of RE shares in the national portfolio. Moreover, they have compared the outcome of various models with the National Renewable Energy Plans (NREPs) [16], showing the strong interaction between RE shares and policy support tools. However, the used models required a level of detail on input parameters that is not easily available for island states. Therefore, in the present analysis a new streamlined optimization tool has been developed that is described in detail in Section 2.3.

1.2. Alternative Development Pathways: Capo Verde, Cyprus and "Barbados Declaration" Islands

The islands that signed the UN on the Small Island Developing States *Barbados Declaration* [17] have ambitious renewable energy targets, aiming to generate a significant share (ranging from 15% up to 100%) of their energy needs using renewables, by 2030. Mauritius committed to increase the share of RES at least to the level of 35% by 2025, developing solar, wind, hydropower, bagasse and landfill gas resources. Seychelles committed to produce 15% of the total electricity from RES by 2030. Amongst the other islands, it is interesting to note that Maldives committed to achieve carbon neutrality in the energy sector and the Marshall Islands plans to electrify all urban households and 95% of rural households using RES by 2015.

Capo Verde has made strong commitments under various schemes, aiming at 100% electricity production from RES [18]. In its NREP, Capo Verde sets the targets to promote the development of renewable energy projects: 25% renewable energy penetration by 2012 and 50% in 2020. It is important that in one of Capo Verde's islands (Brava) the 2020 target predicts 100% electricity production from renewables.

Cyprus also intends to move from the heavy oil and diesel dominance to a sustainable electricity generation portfolio, by boosting energy efficiency and promoting distributed solar energy supply.

2. ISLAND STATES' ENERGY PORTFOLIOS: SUSTAINABILITY ASSESSMENT

The typical strategies that have been used to electrify island states are questionable from sustainability point of view. These strategies often include the installation of a small number of fossil fuel based electricity generation units with the islands' households connected to a centralised grid. Such a scheme has disproportional setbacks in the islands' setting. The fuel has to be transported to the island and stored, therefore, most often they rely on heavy oil and/or coal; the two most carbon emitting energy sources. Long-haul transport makes the fuel more expensive than on continental areas and islands' energy production is very vulnerable and dependant on external resources and higher environmental risk due to oil spill impacts. Oil spillages cause serious harm because of the way they spread [19].

The present study assesses the sustainability of islands' energy systems. This involves a portfolio-optimisation process that aims at analyzing the cost of the long-term operation of the energy system on islands. We have considered an electricity generation portfolio consisting of heavy-oil/diesel fuelled power plants and solar PV systems that can be combined with battery systems to manage the solar electricity between day and night. The indicated battery capacity will be divided between the de-centralised grids, and in the central grid the battery capacities will be distributed to the lowest sub-station levels enabling grid operators to balance between the end-consumers. The selection of these energy sources has been based on the fact that heavy-oil/diesel is currently the most dominant fuel in most islands. At the same time the PV-battery combination has been selected based on the suggestions of the current literature [4], indicating that this combination can be economically advantageous in the concerned islands. Besides, the selected island states have an excellent solar-PV potential (1500–1700 kWh/kWp [20]) which is still untapped (Table 1). Obviously, there are additional options, especially for RES (e.g. wind, hydro, biomass, geothermal). The authors plan to investigate resources of other RES in their future research taking also into account possible synergies through sophisticated system design and hybrid operations [21,22,23].

2.1. Anticipated Evolution in the Electric Power Consumption of the Island States

The developed model in the present study simulates the future electricity demand in island states. Among the model's input parameters are the current installed energy capacity, population, per capita consumption and trends on the consumption increase. Six islands have been selected as case studies of the developed methodology (Papua New Guinea, Capo Verde, Haiti, Malta, Bahamas and Cuba). These island states represent four different regions: Caribbean, South-Eastern Asia, Africa and the Mediterranean and information about them has been included in Table 1.

Table 1. Information of the state islands included in the analysis.

Island State	Population	GDP ($ p.c.)	Consumption (kWh p.c.)	Trend	Fossil (MW)	Solar (MW)	Batt. Storage (MWh)
Bahamas	343,000	22,312	5493	+0.5%	493	0	0
Malta	413,000	22,775	4423	+0.5%	572	2	0
Cuba	11,300,000	6100	1348	+3.0%	5500	65	0
Capo Verde	496,000	3767	580	+5.0%	111	34	0
Papua N.Guinea	6,900,000	2103	486	+5.0%	722	294	0
Haiti	10,000,000	819	36	+5.0%	240	62	0

The island states of Table 1 can be distinguished in three categories, according to their economic status and GDP. Trend column shows the presumed annual increase of the electricity consumption for each country. In Section 1.1 it was shown that electricity consumption is highly related to GDP. Therefore, it is expected that electricity demand on each of the island states will evolve in a different manner, according to the current economic status, the energy trends, population growth rates and future developments. The "consumers" numbers are larger than the population number in Table 1 due to tourism. In the present research the modeling took into account the total consumption (kWh) trends, also including the consumption of the tourism sector.

In the present research we have considered this relation and defined three different scenarios of evolution of the per capita consumption. The first scenario refers to developed island states (in our case Bahamas and Malta) and based on the fact that they are tourist destinations, it predicts a small annual increase of the electricity consumption, equal to 0.5%.

In Cuba the average annual electricity consumption is currently 1348 kWh per person (a low value, especially compared to Bahamas and Malta). Therefore, it is expected that demand will increase in a higher rate, also to respond to the needs of the developing economy. This expectation corresponds to the second scenario with an increase rate equal to 3%.

The current electricity consumption in the remaining islands is extremely low and especially in the case of Haiti is practically zero (36 kWh/capita/year.), indicating the energy poverty of the island. Assuming that the recent economic growth in these island states will continue and will also bring investments on energy infrastructure, a proportionally higher increase on the consumption is expected. For this reason, in the third scenario we have assumed an annual increase of 5% on the annual per capita consumption for Capo Verde, Papua New Guinea and Haiti.

2.2. Optimization Process

The input information of the island states have been utilized by an optimization model that has been developed for the present study. This models builds on a modern metaheuristic algorithm, Harmony Search [24]. The specific algorithm has been already successfully applied to solve various engineering problems, including application on energy production [25,26] and systems' energy efficiency [27].

In the present study the developed algorithm performed independent runs for each island state, aiming to estimate the future status of the electricity demand and the required investments on new energy capacity. Following the analysis of the input information for each island and the specific trends on energy consumption and population growth, the developed model performed an optimization on the future additional capacity installations, on a 20-year horizon. This is the core of the present analysis, the model examining the different options in each island state in order to address future (2015–2035) electricity needs, in the most efficient manner. For each island numerous runs have been performed and each run searched the solution-space for the most advantageous (global optimum) solutions. For each island state, the model eventually converged and suggested solutions that correspond to energy investment options with the best value of the objective function (lowest cost) that also satisfies the various constraints.

2.3. Model Formulation: Decision Variables, Objective Function and Constraints

2.3.1. Decision Variables

The model's decision variables correspond to additional energy capacity that will be required in each island state in the studied time frame (2015–2035). This includes investments in solar PV systems (variable X1) and traditional fossil fuel based systems (variable X2). Moreover, the model investigates the possibility for battery storage (variable X3) as well as the operation hours of fossil systems in an annual basis (variable X4). Initial values of the decision variables correspond to the current status, namely the characteristics of existing systems.

2.3.2. Constraints

Obviously, the basic constraint of the model is that the proposed portfolio will manage to cover the electricity demand, throughout the studied period. It is important to note that the electricity production needs also to cover energy losses in the transport and distribution networks as well as losses due to any other reason. Transmission and distribution losses in the selected island states range from $\simeq$11% in Malta to $\simeq$20% in Papua New Guinea ([13,14]). Accordingly, the developed model considered a uniform value of 15% on top of the consumed electricity, in order to estimate the required electricity production. It is important to underline the large losses of Haiti's network that reach values as high as 55%. Haiti is a unique case and as presented in Section 3.3, energy poverty is a major challenge in this island state.

2.3.3. Objective Function

The objective function of the optimization process calculates the total cost of the suggested energy portfolio (installation and operation cost) for the studied period (2015–2035).

$$Total\ cost(j) = Capital\ cost(j) + Operation\ cost(j) + CO_2\ tax(j) \tag{1}$$

where

- j: year under study
- Capitalcost: cost of new installed capacity (PV, fossil, battery storage)
- Operationcost: fuel cost, CO2tax: carbon tax

Incoming and outgoing cash flows of Equation (1) over the period of time have been transformed into net present values at a discount rate equal to 3%. The efficiency of the different systems, capacity factors as well as the installation costs have been based both on the previous work by the authors [15] and the recent developments reported in [3,4].

Energy Systems' Capital Cost

In October 2014, the worldwide average price of a residential PV system without tax was given as €1.45/Wp (about 15% higher than in Europe with €1.27/Wp) [8]. This cost, illustrated in Figure 2, was used by the model as the systems' PV cost. The similar input cost figures for the fossil fuel capacities were €0.9 million/MW and for the battery storage €0.12/MWh.

Fossil Fuel Price and CO2 Tax

The high price of diesel and heavy oil fuel used for power generation in the island states results in commercial electricity tariffs that are much higher than the global average. In 2012 the average electricity price in Caribbean islands and Mauritius was €0.3/kWh, while in Hawaii it reached €0.395/kWh. Cyprus and Malta have the highest electricity prices (excluding taxes) in the European Union (EU).

The fluctuation of oil and gas prices makes it a complex task to compare the economic sustainability or the cost-effectiveness of the potential electricity generation technologies. From energy management point of view diesel becomes inefficient if not operated at full load. For this reason the developed model also monitors the duration of the fossil fuel systems' operation for each island state.

The analytical model was executed using two different fuel-price scenarios. The first scenario builds on the 2012 heavy oil/diesel component of the electricity production in Cyprus with a cost of €0.121/kWh. Since 2012 the market has experienced roughly 50% decrease, that considerably affected the energy landscape. For this reason we have added a second scenario where the electricity price is €0.06/kWh and predicts an annual increase of +2.5% (a conservative increase). In order to keep the present paper concise, the results-graphs presented in Section 3 are limited to the second more up-to-date scenario. Analysis and results of the high fuel-price scenario are available with the online material.

The cost of fossil fuel electricity production was also attributed an additional CO2 tax, in order to anticipate the developments of the electricity market in the following decades. The input values for capital and operation costs are presented in detail in Table A1, in the Appendix.

Time Horizon of the Simulation

The present analysis has a 20-year horizon and examines the different options for the island states for the period 2015–2035. However, it is important to include costs of the years beyond the studied period in order to take into account the depreciation of investments made near the end of the studied period. In case this particularity is ignored, the optimization process will be biased and converge to large heavy fuel installations for the years just before 2035, since future fuel costs will not be taken into account.

This particularity has also appeared in a previous study by the authors [15] and was addressed by projecting the optimization process in a longer future time-frame. Similarly, in order to acquire non-biased results and avoid manipulation of the model, the optimization process was set to run for the period 2015–2055, prolonging further in the future for 20 years more. In that way, the results until the year 2035 also consider costs that will appear in the future from investments made before 2035.

In order to take the aging of PV modules also into account, an annual decrease of 0.5% for each year of operation has been introduced for installed solar PV systems. This value has been based from field testing results of last 40 years, as published in the literature [28].

The analysis also considered the fossil fuel stations' decommissioning. Most of the islands under study have aged fossil fuel stations that will either be phased out or retrofitted. This will require additional capacity investments that are indicated by "spikes" in electricity production (Figure 7, Figure 8, Figure 9, Figure 10, Figure 11).

Battery Storage

The developed optimization model on one hand converges to the best solution that defines values of the required annual additions on the capacity of fossil-fuels and solar-PV. Moreover, it also examines the extent on which battery storage is advantageous. In order to achieve that, it distinguishes day from night electricity consumption. A simple analysis of the studied islands showed that it is relatively accurate to assume 2/3 of the total demand (5840 h/year) corresponding to the day consumption and 1/3 (2920 h/year) to the night consumption. This assumption has been applied in a uniform manner to all islands in order to keep the model as simple and efficient as possible.

Another assumption considers that battery storage and its operation aims to cover exclusively night demand. Thus, batteries will be designed to consume part of the produced solar-PV electricity during the day to recharge and provide the stored electricity at night hours. In that way batteries provide electricity PV systems are out, fossil fuels cover peak-loads and support the night operation, while PV systems minimise or eliminate the oil/diesel consumption. Battery systems can also provide flexibility to the system's operation during the day, through temporal electricity storage, provided that they will be fully charged before their night operation starts. Islands with bigger reservoirs could also include pumping-storage for balancing reasons (however the present analysis does not cover this aspect). This service has an indirect value and might be very important in some cases. However, the present analysis considers it as a technical advantage and has not assigned an economic value to it that should be included in the cost function.

3. RESULTS

In the present section the results for the six island states are presented (Papua New Guinea, Capo Verde, Haiti, Malta, Bahamas and Cuba). As explained in Section 2.3 the optimization algorithm performed independent runs for each country. Initially information on production/consumption patterns of each island state was fed into the algorithm. Then, the metaheuristic algorithm performed the iterative optimization process aiming to minimize the objective function (Equation (1)) for the system's overall operation period. Thus, the algorithm explored different combinations of capacity additions, energy storage and fossil fuel consumption that constitute the optimum option. The algorithm also considered aging and phase out of the energy systems.

Eventually the algorithm converged for each island state to the annual values of the four variables (a set of 20 values of X1−X4 for the period 2015–2035) that minimize the total cost. These values are illustrated in Figure 6, Figure 7, Figure 8,Figure 9, Figure 10, Figure 11 and represent the low-cost option for the island states under study.

As explained in Section 2.3.2, the optimization model aimed at an electricity production that also covers the various electricity losses and it is higher from the consumption. This additional electricity production is conspicuous in some of the graphs (Figure 8 and Figure 11), but less visible in others (Figure 8 and Figure 11). Indeed for several years the optimal system configuration predicts a production equal to the consumption where the two lines production-consumption coincide (e.g., Haiti between the years 2020–2024).

In the output figures there is an unanticipated peak of production at the midperiod for all islands. The reason behind this is that model assumed 10 years of remaining lifetime for the existing fossil capacities. That appeared to be the case for at least one of few units for the islands and causes an overcapacity accumulation just before the replacement.

As described in Section 2.3.3 the model optimizes the values of capacity additions for energy systems that will cover the future demand. As a result the spike in Figure 6, Figure 7, Figure 8, Figure 9, Figure 10, Figure 11 indicates a hypothetical electricity production that would appear only if old and new capacities continued to produce electricity in the same manner. Electricity production line is a projection of the optimal capacity additions. Obviously, in real world this production spike would involve large, unreasonable energy waste. Therefore it is expected that capacities that will have the higher electricity production cost will remain idle in the year of the spike until older units will be decommissioned.

3.1. Model Results for Papua New Guinea

Papua New Guinea represents the Pacific islands, where 70% of households do not have access to electricity. In fact, Papua New Guinea together with the Solomon Islands and Vanuatu have electrification rates that are among the lowest in the world (<20%). Such values can only be compared to the poorest African countries. In the projected 10 year period, the island's predicted electricity consumption (Figure 6) shows a threefold increase due to its very low starting point (extremely small per capita consumption) and also due to the dynamic population increase. Among the islands studied in the present research Papua New Guinea is a medium populated but the population is increasing rapidly.

The results of the simulation showed that the best energy strategies for 2015–2035 suggest a dynamic PV installation in the beginning of the period with increase in the battery capacity. Heavy oil capacities are increased towards the end of the 20-year period, following their increased operations in both the daytime and night. With the (assumed) phase out of the original oil capacity the nominal PV capacities exceed the fossil capacities by 50%.

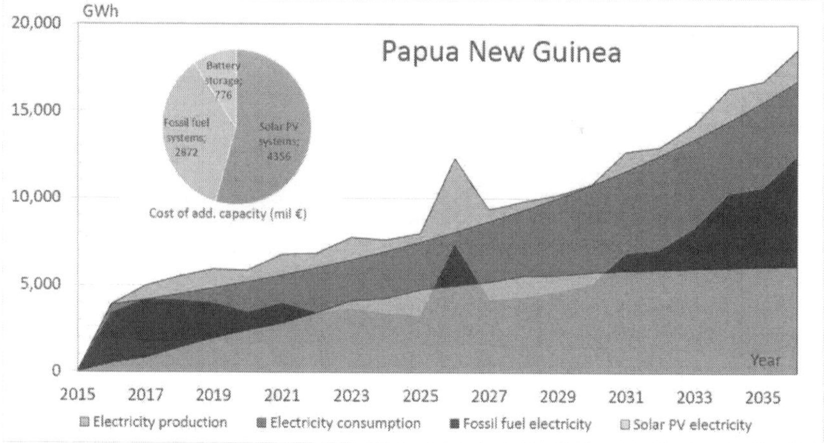

Figure 6. Papua New Guinea: Additional electricity production (optimal proj.: 2015–2035).

3.2. Model Results for Capo Verde

The West African island state has a small population compared to the other islands included in the analysis. It also has a very low per capita (p.c.)

electricity consumption. It is part of the Economic Community of West African States (ECOWAS) and hosts the headquarters of the ECOWAS Regional Centre for Renewable Energy and Energy Efficiency (ECREEE). Though Capo Verde islands are characterised by low electrification rates, the national government has set ambitious energy plans with the support of the ECOWAS-ECREEE, aiming at an eventual 100% RES electricity production. Accordingly, several renewable energy projects that rely on solar and wind distributed generation are already on the construction/operation phase.

The optimization model supports the adequacy of the local RES potential and the soundness of the government's plans to rely on renewables (Figure 7). It projects that even in a low oil-price scenario the PV capacity should represent almost 75% of the energy portfolio with the fossil fuel capacities hardly contributing to the daytime load of the islands.

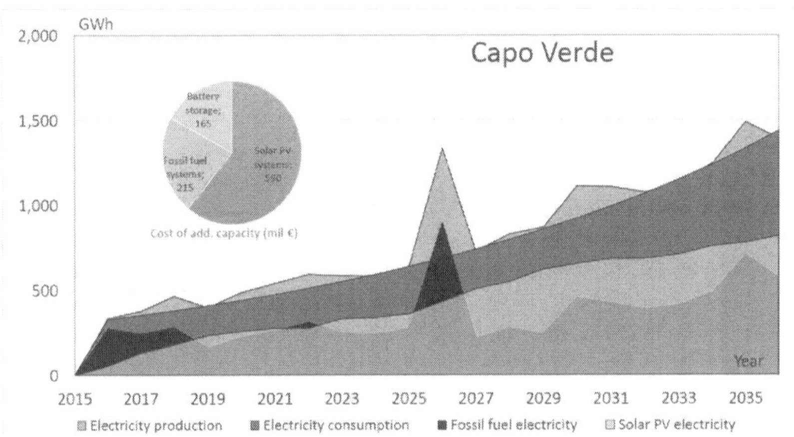

Figure 7. Capo Verde: Additional electricity production (optimal projection: 2015–2035).

3.3. Model Results for Haiti

Haiti, together with Cuba represent the largest islands selected for the present analysis and also the poorest in the group (Figure 8). Despite being a large island, its electricity consumption is more similar with the smallest islands discussed here. Thus, the electrification and consumption rates are amongst the lowest in the world. Energy poverty will need to be mitigated under any future sustainable development scenario. Accordingly, the model assumed the highest increase on the electricity consumption for Haiti.

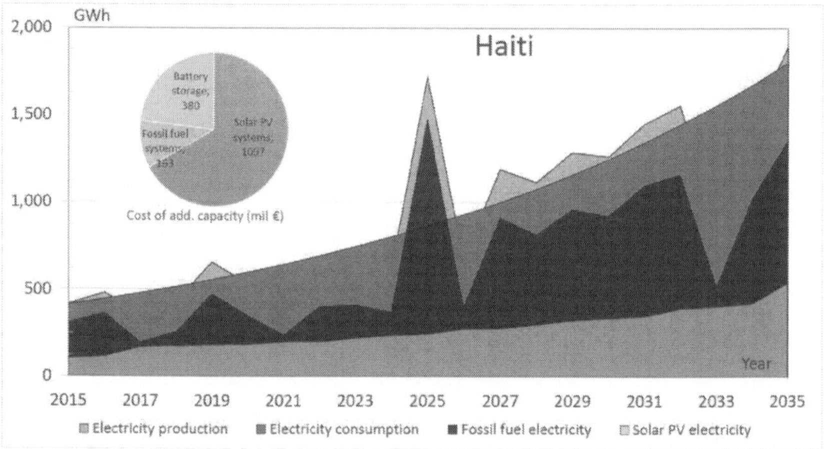

Figure 8. Haiti: Additional electricity production (optimal projection: 2015–2035).

In order to address this dynamic consumption increase, the optimization model's output suggested the highest annual fossil fuel capacity additions among the analyzed islands. The present situation justifies an accelerated capacity installation pathway, which renders as secondary the technology choice. Nevertheless, the balance PV and oil capacity shares in the first 10 years will provide time and experience to examine whether a more sustainable pathway could be feasible in the second half of the period, perhaps taking advantage of the financial and fiscal instruments highlighted in the conclusions.

3.4. Model Results for Malta

Malta represents a small island that is part of the European Union, with high values both of GDP and electricity consumption per capita. At the same time, Malta has a very low share of RES. For the reasons discussed in Section 2.1, a more moderate consumption and population increase is foreseen for Malta. A 100 km interconnector power line with Sicily is planned and is expected to be operational by the end of 2015. The interconnector power line will increase the available capacity by 200 MW. The optimization model converges to high PV penetration for Malta, although the initial consumption is high and as a result the expected capacity installations are more moderate (Figure 9). Fossil fuel based generation stays at a near constant level (with the exception of the assumed phase-out year) since the total production is expected to remain at a quite high level. The dynamic PV additions show that the PV option offers a cost-effective way to diversify the electricity generation from one dominant resource to a developed island state. This could help to achieve not only a more sustainable energy mix, but move the country towards energy security.

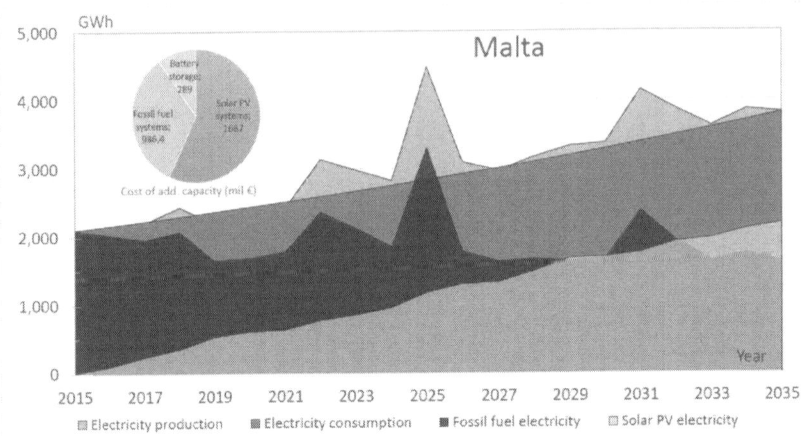

Figure 9. Malta: Additional electricity production (optimal projection: 2015–2035).

3.5. Model Results for Bahamas

Bahamas is the richest amongst the six islands, having the highest electricity consumption level. The optimization resulted in the highest fossil plant utilization, but towards the end of the studied period a convergence to nearly equal share between PV and fossil based capacities was observed (Figure 10). This clearly shows the potential of PV capacities to support additional electricity demands caused by the expansion of tourism, even in island countries that already have a well-developed power portfolio.

3.6. Model Results for Cuba

Cuba has the biggest population and medium GDP and consumption per capita amongst the selected islands. There are several socio-economic differences to the other five examples that could not have been taken into account in the input parameters. These differences make it difficult to make recommendations on incentive schemes.

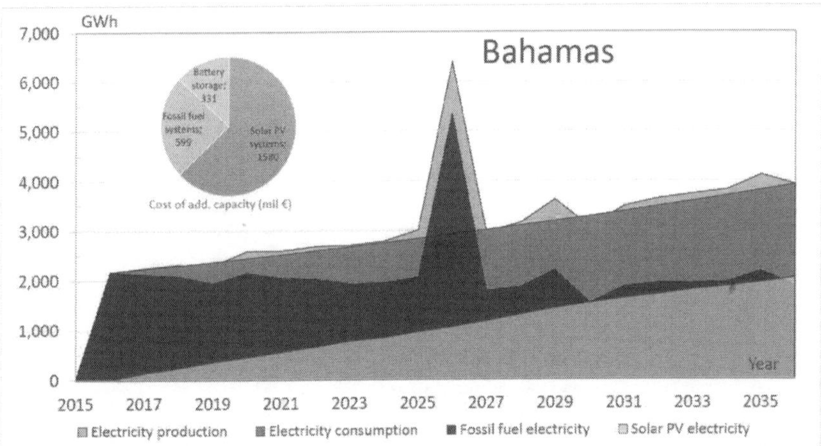

Figure 10. Bahamas: Additional electricity production (optimal projection: 2015–2035).

However, the optimization exercise was useful to highlight the good potential for this size of islands as well. The model converged to PV and fossil capacities reaching 40%–60% in the electricity production by the end of the study period, according to the fuel-price scenario. The model's output gives a better insight to the robustness of Cuba's original electricity mix and highlights the significant huge potential for a RES market penetration in a good geographic location with socio-economic particularities.

There are quite a few factors that could help the Renewable Energy development in Cuba, that are discussed in the following section. In the near future Cuba will need to take significant steps in converting its energy portfolio to a more sustainable one that will rely more heavily on the indigenous energy sources. The rural areas of the country have during the last two decades implemented many projects with international cooperation, that are based on decentralised energy systems. The electricity grid infrastructure in Cuba will require major refurbishments which could also be advantageous for distributed energy options.

3.7. Numerical Summary of the Results

The resulting share of PV, battery storage and fossil fuel capacities as well as an estimation of the electricity generation for the studied island states is illustrated in the following Table 2.

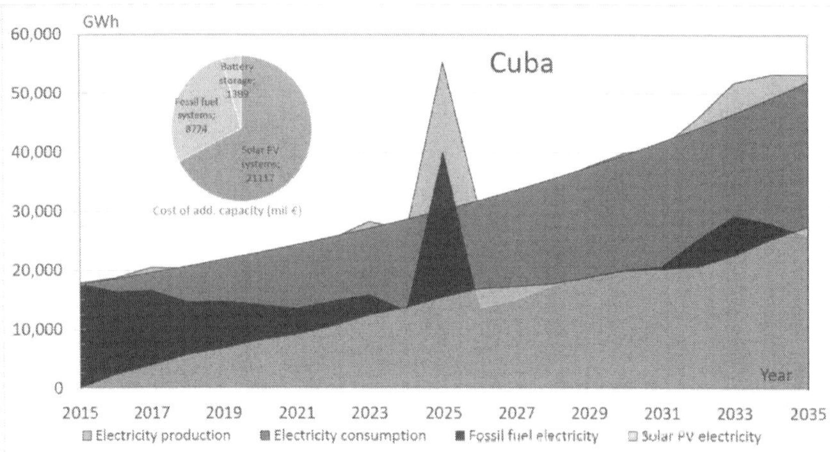

Figure 11. Cuba: Additional electricity production (optimal projection: 2015–2035).

Table 2. Resulting share of Fossil fuel, PV and battery storage systems.

PV, Fossil Fuel: MW Battery Storage: MWh	Papua New Guinea	Capo Verde	Haiti	Malta	Bahamas	Cuba
Fossil Fuel (existing)	722	111	240	572	493	5500
PV (existing)	294	34	62	2	0	65
Battery storage (existing)	0	0	0	0	0	0
PV percentage % (existing)	41	31	26	0	0	1
Fossil Fuel (in pipeline)	530	31	0	0	0	370
PV (in pipeline)	0	349.8	0	0	0	0
Fossil Fuel (modelled–2035)	226.5	272	528	538	568	797.1
PV (modelled–2035)	3273	698	321	1330	863	1194.3
Battery storage (modelled–2035)	1458	402	1162	585	815	2132
PV percentage % (modelled–2035)	145	257	61	247	152	150

4. DISCUSSION AND CONCLUSIONS

Optimizing the energy portfolios for the islands to the minimum cost level means a move towards providing electricity by more sustainable resource base. The resulting electricity generation portfolios for all studied islands clearly increased the use of indigenous sources for electricity generation compared to the current production dominated by fossil fuel.

Although in most cases the heavy oil capacities remain in the portfolio at a constant or even increased capacity level, its share and daily utilization decreases as the PV capacities are projected to build up in the cost optimal portfolio. This means that solar resource starts to substitute the heavy oil/diesel use, decreasing the fossil fuel shipment-storage cost, CO2emissions, environmental risks as well as the dependence on imported energy resources. And this movement towards a more sustainable and indigenous resource-use secures at the same time cost savings even at the conservative fossil fuel cost scenario.

A significant tendency, a lock-in mechanism can be observed on the resulting cost optimal power portfolios. Islands that initially have high shares of fossil fuel capacities are expected to keep these higher shares at 2035 (e.g., Figure 6). Therefore, it appears the existence of RE systems in the islands facilitates further development and vice versa.

Immediate PV installations of many small capacities at the beginning of the projected period appears as a beneficial feature for further PV additions (modular design). In that way investors instead of funding a single fossil power plant invest small instalments on modular projects. Risk is thus reduced and solar PV portfolio can be gradually developed, when advantageous. This gradual transition has the extra advantage that mitigates the lock-in effect that early installations of large fossil fuel capacities have: price fluctuations might render them idle for long periods, preventing at the same time investments on RE. Recent is the example of the gas turbines in Europe with large capacities remaining idle since 2012.

International donor organisations and agencies specialised in energy sector have to play a more active role in influencing the National Electricity Master-plans of islands. The present plans are dominated by fossil based power generation options that are not optimal: neither from sustainability nor from cost optimal solution point of view. Recent donor supports for island states have to be reviewed and should eliminate subsidies for resources with harmful climate effects.

The gradually increasing development of the PV sites could also delay the decision on the phase out of existing large fossil fuel based capacities. By the time this phase out would take place the other clean technologies based on indigenous resources (like the proposed large-scale hydroelectric projects in Papua New Guinea, Jamaica and Guyana or the wind potential in many islands) would be able to completely substitute the fossil fuel based ones.

AUTHOR CONTRIBUTIONS

Sándor Szabó conceived the object of the research. He identified and collected the required input information and wrote most of Section 1, Section 3 and the conclusions. He supervised the research. Ioannis

Kougias developed the optimization model and wrote the code in MATLAB. He performed the runs and collected the information. He also wrote the relevantSection 2. Magda Moner-Girona analyzed the outcome and suggested the instruments to support RES deployment in island states. Katalin Bódis performed exploratory analysis on the collected data and was responsible for the visualization and reporting. All authors have read and approved the final manuscript.

CONFLICTS OF INTEREST

The authors declare no conflict of interest.

DISCLAIMER

The views expressed in this paper are purely those of the writers and may not in any circumstances be regarded as stating an official position of the European Commission.

APPENDIX

Table A1. Cost values used in the modeling process.

	Fossil Fuel Systems	Solar PV Systems	Battery Storage
Capital Cost	€900,000/MW	€1,300,000/MW	€0.12/MWh
Fuel price (1st scenario)	€121/MWh	-	-
Fuel price (2nd scenario)	€60/MWh + 2.5%/year	-	-
CO_2 tax	€11/MWh	€0/MWh	€0/MWh

REFERENCES

1 United Nations General Assembly. *World Summit Outcome*; Sixtieth Session; United Nations Publication: New York, NY, USA, 2005.

2 Intergovernmental Panel on Climate Change (IPCC). *Climate Change 2013: The Physical Science Basis. Contribution of Working Group I to the 5th Assessment Report of the IPCC*; IPCC: Geneva, Switzerland, 2013.

3 GlobalData: Global research and consulting for the energy and healthcare industries. Available online: http://www.globaldata.com/ (accessed on 9 February 2015).

4 International Renewable Energy Agency (IRENA). *Renewable Energy Country Profiles*; Special Edition on Islands; IRENA: Masdar City, UAE, 2012.

5 Eurostat. *Statistics Database*; Eurostat: Luxembourg, 2013.

6 Szabó, S.; Bódis, K.; Huld, T.; Moner-Girona, M. Energy solutions in rural Africa: Mapping electrification costs of distributed solar and diesel generation versus grid extension. *Environ. Res. Lett.* **2011**.

7 Szabó, S.; Bódis, K.; Huld, T.; Moner-Girona, M. Sustainable energy planning: Leapfrogging the energy poverty gap in Africa. *Renew. Sustain. Energy Rev.* **2013**, *28*, 500–509.

8 Jäger-Waldau, A. *PV Status Report 2014*; Publications Office of the EU: Luxembourg, 2014.

9 Taylor, M. The Renewable Revolution: Wind Power Costs, Visual Presentation. November 2012. Available online: http:// www. slideshare. net/MichaelTaylor6/wind-power-costing-27-nov-2012-at-the-energy-talk-london (accessed on 13 July 2015).

10 Vansintjan, D. The energy transition to energy democracy. In *Report of the REScoop 20-20-20 Intelligent Energy Europe Project*; De Wrikker: Antwerp, Belgium, 2015.

11 Nfah, E.M.; Ngundam, J.M. Feasibility of pico-hydro and photovoltaic hybrid power systems for remote villages in Cameroon. *Renew. Energy* **2009**, *34*, 1445–1450.

12 Bekele, G.; Tadesse, G. Feasibility study of small Hydro/PV/Wind hybrid system for off-grid rural electrification in Ethiopia. *Appl. Energy* **2012**, *97*, 5–15.

13 Energy Information Administration (EIA). *Annual Energy Outlook 2015*; Office of Communications: Washington, DC, USA, 2015.

14 IEA; OPEC; OECD; World Bank. Analysis of the Scope of Energy Subsidies and Suggestions for the G-20 Initiative. Available online: http://www.oecd.org/env/45575666.pdf (accessed on 1 September 2015).

15 Szabó, S.; Jäger-Waldau, A. More competition: Threat or chance for financing renewable electricity? *Energy Policy***2008**, *3*, 1436–1447.

16 Szabó, S.; Jäger-Waldau, A.; Szabó, M.; Monforti-Ferrario, F.; Szabó, L.; Ossenbrink, H. European renewable government policies versus model predictions. *Energy Strategy Rev.* **2014**, *2*, 257–264.

17 United Nations Developing Programme. Barbados Declaration on Achieving Sustainable Energy for All in Small Island Developing States (SIDS). In *Barbados Declaration*; United Nations publication: New York, NY, USA, 2012; Available online: http://www.undp.org/content/dam/undp/library/Environment%20and%20Energy/Climate%20Change/Barbados-Declaration-2012.pdf (accessed on 1 September 2015).

18 United Nations Industrial Development Organization (UNIDO); Regional Centre for Renewable Energy and Energy Efficiency (ECREEE). *Cape Verde: Energy Analysis and Recommendation Undertaken by UNIDO and ECREEE Report*; ECREEE: Praia, Cape Verde, 2011.

19 Wang, H.; Xu, J.; Zhao, W.; Zhang, J. Effects and Risk Evaluation of Oil Spillage in the Sea Areas of Changxing Island.*Int. J. Environ. Res. Public Health* **2014**, *11*, 8491–8507.

20 Šúri, M.; Huld, T.A.; Dunlop, E.D. PVGIS: A web-based solar radiation database for the calculation of PV potential in Europe. *Int. J. Sustain. Energy* **2005**, *24*, 55–67.

21 Kougias, I.; Bódis, K.; Jäger-Waldau, A.; Szabó, S. Identifying the Energy Potential from PV Installations on the Face of Dams in Mediterranean Islands. In Proceedings of the EU PVSEC International Conference, Amsterdam, The Netherlands, 22–26 September 2014.

22 Kougias, I.; Bódis, K.; Jäger-Waldau, A.; Monforti, F.; Szabó, S. Exploiting existing dams for solar PV system installations. *Prog. Photovolt.: Res. Appl.* **2015**.

23 Monforti, F.; Huld, T.; Bódis, K.; Vitali, L.; D'Isidoro, M.; Lacal-Arántegui, R. Assessing complementarity of wind and solar resources for energy production in Italy. A Monte Carlo approach. *Renew. Energy* **2014**, *63*, 576–586.

24 Kougias, I.; Theodossiou, N. A new music-inspired harmony based optimization algorithm. Application in water resources management problems. In Proceedings of the PRE X International Conference, Corfu, Greece, 5–9 July 2010.

25 Salcedo-Sanz, S.; Pastor-Sánchez, A.; Del Ser, J.; Prieto, L.; Geem, Z.W. A Coral Reefs Optimization algorithm with Harmony Search

operators for accurate wind speed prediction. *Renew. Energy* **2015**, *75*, 93–101.

26 Kougias, I.; Theodossiou, N. Application of the Harmony Search optimization algorithm for the solution of the multiple dam system scheduling. *Optim. Eng.* **2013**, *14*, 331–344.

27 Kougias, I.; Theodossiou, N. Multiobjective pump scheduling optimization using harmony search algorithm (HSA) and polyphonic HSA. *Water Resour. Manag.* **2013**, *27*, 1249–1261.

28 Jordan, D.C.; Kurtz, S.R. Photovoltaic degradation rates-an analytical review. *Prog. Photovol. Res. Appl.* **2013**, *21*, 12–29.

Index

W